KU-113-436

THE HIDDEN WORLD OF THE FOX

THE HIDDEN WORLD OF THE FOX

ADELE BRAND

William Collins
An imprint of HarperCollins*Publishers*
1 London Bridge Street
London SE1 9GF

WilliamCollinsBooks.com

First published in Great Britain in 2019 by William Collins

1

Artwork by Jo Walker

A catalogue record for this book is available from the British Library

ISBN 978-0-00-832728-6

Printed and bound by CPI Group (UK) Ltd, Croydon, CR0 4YY

This book is produced from independently certified FSC™ paper
to ensure responsible forest management

Find out more about HarperCollins and the environment at
www.harpercollins.co.uk/green

For my parents

CONTENTS

1 Who is the Fox? 1

2 A Brief History of the Fox 10

3 Where do Foxes Live? 29

4 What Does the Fox Look Like? 43

5 Fox Family Matters 56

6 The Fox and its Neighbours 74

7 What Does the Fox Say? 94

8 Counting Foxes 110

9 In Sickness and in Health 121

10 Predators Among Us 138

11 When the Fur Flies 151

12 Tornado in a Cage 163

Epilogue: Fame and Foxes 177

The Fox Watcher's Toolkit 183

Acknowledgements 196

Bibliography 197

Image Credits 216

1

Who is the Fox?

VISUALISE A FOX: flame-orange on a white canvas, black paws and thick brush, pointed muzzle and diamond-sharp eyes. Now paint its native wildwood behind it – this fox is trotting through the undergrowth, exploiting trails within the brambles trampled by badgers. It leaves neat narrow tracks on mud softened by afternoon rain, and snags its fur on thorns in passing.

Woodland, farmland, hedgerows and weary old trees. Owls, hedgehogs, rutting deer. Dead man's fingers – that is, grisly-looking black fungi – poking through sweet chestnut leaves in the autumn; woodpeckers playing rat-a-tat-tat on dying branches in the spring.

This is the classic British landscape of the classic British fox: the precious fragments of countryside saved from industrialised agriculture and overdevelopment. Ancient, intriguing, revitalising and poetic, our rural semi-wild has enchanted animal-focused authors from Beatrix Potter to Colin Dann of *The Animals of Farthing Wood* fame. The fox of tradition lives squarely within it, running under a cloud of mythology stirred by friend and foe alike.

But it is not the only fox of twenty-first-century Britain.

IMAGINE ANOTHER DUSK, this one after a day when chainsaws groaned and concrete mixers churned, and builders wolf-whistled at local women from a half-built rooftop. Woodland here is being transformed into a housing estate, rimmed by a newly built wall thick enough to please Hadrian, its bricks highlighted in passing by the headlights of commuter traffic.

A small vixen with a slender face and wary eyes tugs at chips dropped by the workmen, slicing artificially flavoured potatoes with enlarged molars called carnassials which define her species as a member of the Carnivora. She digs under the perimeter fence, and darts across the main road, feet fast and brush bouncing, passing me as I walk my dog. Ironic, perhaps, for wolves – the ancestor of dogs – once lived here too, feeding foxes through scraps of deer meat. The last Home Counties wolf was killed in Hampshire 800 years ago. The crowds returning from

London have forgotten; perhaps the woodland has not. In an ecosystem, every extinction is like snapping a link in a chain.

But foxes themselves are in no danger of disappearing. Into a driveway the vixen turns, past trees native to China, through a side gate sealed against burglars, into a garden where another fox is burying Bakers Complete dog biscuits. The little vixen is an intruder in this territory. The resident flies at her, flipping her upside down, and skull-splitting screams – theatrical, but bloodless – pepper the night over the droning of the traffic.

She struggles free, and bolts back across the road into the fragmented woodland. Her motive for this daring if ill-fated trespass is obvious: she is lactating and needs food and water to produce milk for her cubs. She is driven by an unquenchable instinct to survive.

THAT WAS LAST YEAR'S DRAMA.

I haven't seen that particular vixen for a few days; it is mid-March as I write this, and doubtless she is underground with a new litter. She has survived the last twelve months despite her wood being turned into houses with million-pound price tags, and despite the best efforts of the neighbouring fox group to keep her out of the garden. Her body language is tenser than theirs, her eyes a little sharper, and her habit of poking her muzzle through gaps in the fence never fails to amuse.

This is not the city; it is Surrey's battered greenbelt. Despite the developers stalking the county like thieves eyeing up wallets, we still have rich and abundant wildlife between the golf courses, out-of-town supermarkets and ever slower M25. Yet only a few miles north of the endangered wildflowers thriving on our chalky hills, the mood changes. London town spikes our northern horizon with towers, giant wheels and an orange nocturnal haze. Somehow, once there, we consider it unremarkable that we have grown buildings taller than trees.

It is undeniably beautiful, that old city filled with lion statues. History smiles from every spire and road name, grand, grotesque or tragic. You fall into the rhythm of it: the river of people flowing from Victoria station in the mornings, the shouts of *Big Issue* sellers, tourists photographing themselves in St James's Park. Cyclists speeding across pelican crossings, strangers apologising in the street when you bump into them, anti-war protesters perched on window frames with placards while weary police keep watch – it is such a human place.

Human, but full of foxes. Many thousands of them live in urban environments in Britain, from London to Edinburgh.

We jolt at that, sometimes alarmed, sometimes happy that a being of the ancient wildwood can find a home in

Britain's sprawling capital – it feels a little out of sync, like an Elizabethan lady in ballroom dress among the revellers in All Bar One. The contrast between free wild animal and hard concrete street is vivid, irresistible, burning a place in our collective consciousness. Fed on television images that associate wildlife with wilderness, this displacement of 'normal' can beget either wonder or fear. Perhaps the social reserve in the British psyche leaves us puzzling over the correct etiquette. Upon seeing a fox, many people are not quite sure what to say.

So, instead, we have put the fox in the dock for questioning. We have accused it of trespassing into the human domain, of being cheeky, of spreading disease, harming pets, and even posing a significant risk to ourselves. Unperturbed, the fox strays ever further into our world, permeating our language, pop music, movies, pub names and television adverts. They are debated in offices, schools and Parliament. One was recently filmed by bemused journalists outside 10 Downing Street as they awaited the appearance of the prime minister. Another found fame climbing 72 floors of the Shard, and lives on in that monolith's merchandise. Others have trotted onto the pitch in the middle of high-profile football matches.

Bizarrely, even our real courtrooms are not immune. Temping as a court officer to staunch debts after my graduation, I was surprised to hear the defendant in my very first criminal trial claim an alibi of being busy feeding a 'baby' fox. She was still found guilty; it is beyond the court's powers to summon foxes as witnesses.

Foxes have filled my life, too; it has become a running joke among friends that wherever I go – from the Indian desert to the Yucatan rainforest – I am bound to meet one, usually sitting, as they do, watching me from a distance. They dominated my wildlife diaries as a child, were part of my academic studies in ecology, and have always been the most popular stars for the millions of visitors dropping by my corner of the internet. I have fostered orphaned cubs and injured adults for the Fox Project charity, and been privileged to observe and film some extraordinary fox behaviour in the wild. Mostly, however, I wish to know them as individuals, to learn the stories of their lives as honest biographer – and to be a mediator, hoping to keep the peace between human and fox.

Through that, I have crossed the trail of two foxes: the wild one which fills my spreadsheets with scientific data, and its non-identical twin that dwells in the human imagination. Twenty years of observing, photographing, and occasionally rescuing foxes have impressed on me just how very complex a neighbour we have in this small, curious member of the dog family. But the human response to wildlife can be just as nuanced. I've seen the extremes of it: the fear, the hate, the passion and kindness.

This matters. The world is now mostly humanised. There may be valleys in Tasmania which have never been explored, and tundra lakes in the great Canadian north that are lonely save for mosquitoes and caribou, but for many wild animals eking a living while you are reading these words, wilderness is irrelevant. They're living on land that is controlled by humanity. From forests heavily managed for commercial timber to grasslands seeded with exotic crops and split by dangerous roads, many creatures must compensate daily for anthropogenic changes to landscapes that they occupied long before palaeontologists revealed the existence of deep time.

Yet this overlap zone, where civilisation and wilderness meet, is not devoid of biodiversity. With tolerance, respect or sometimes by simply ignoring, nature can thrive in the human shadow. Urban wildlife is here to stay, and not only in London. Leopards share the exotic bustle of Mumbai with twelve million people. Spotted hyenas scavenge rubbish in major African cities. Vancouver occasionally debates the pumas that stalk mule deer in suburban gardens. And foxes, no less controversial than the great carnivorans, have adapted to the new biome called 'city' from Aberdeen to Zurich, from the bitter winters of western Canadian metropolises to the scorching desert towns of Israel.

Sharing the same geographical space as wildlife brings out instincts in people that were more proportionate in days when we had to fend off sabre-tooth cats. In a world full of modern dangers, we are haunted by the idea of a primeval fate. The results of that fear can be ugly. I've watched Canadian police officers kill bears that were harming no one because, well, they just couldn't be sure what tomorrow might bring. False widow spiders, coyotes, wolves, raccoons, foxes – they've all had their headlines.

But as night falls in my 1,300-year-old Surrey village, the other side of the equation swings into life. All down these streets are householders who will smile at a fox trotting across their lawn tonight. Fear may have grown as we have become ever more disconnected from nature, but so has a desire to rekindle that relationship. The small glimpse of a wild fox – and, it has to be said, the controversial practice of deliberately feeding them – brings a lot of happiness to many.

MY AIM WITH THIS BOOK is to explore how the red fox, a wild animal that evolved in the wildwood, has adapted with such dramatic success to modern Britain. This involves understanding the real fox as researched by cutting-edge science, and considering its behaviour, physical form and intelligence in the context of the world that it inhabited for thousands of centuries before finding us.

This is not a book about fox hunting. That argument has consumed multitudes of space elsewhere. Once the real creature displaces the mythological fox of hunters' lore, and a vague sense that 'populations must be controlled' is replaced with scientific knowledge, the question of whether arbitrary cruelty is acceptable rather answers itself.

To a small extent, this is also a book about people: how we form our opinions of nature, and why honest observations can sometimes be misleading. To clarify, I am not anti-human. Environmentalists who are, doom themselves to an eternity of digging tunnels for Swampy and being ignored by decision-makers. Education is more effective than alienating the public with abrasive name-calling – a lesson some animal rights activists would do well to remember. Assuming that everyone who is concerned about foxes sharing their garden must be a paid-up member of the Countryside Alliance is about as realistic as Brer Fox designing a Tar Baby.

This great British public, these people whose world overlaps that of foxes – they are binmen, bankers, the bankrupt, golfers, mothers caught in traffic on the school run, even criminals.

This is you, England.

You are beautiful, heart-breaking, eccentric and implausible.

You are the people who foxes tolerate as neighbours.

The question is now, will you tolerate them?

2

A Brief History of the Fox

WE ARE REDESIGNING THE FOX: its diet, territory size, social interactions, and its longevity and causes of death have all been changed by us. Even their body fats are impregnated with our lifestyle, carrying residues as diverse as fire retardants and nuclear radiation. Their days are filled with human-made noises, human-made landscapes, and human-made risks.

But foxes have not spent their evolutionary history sunbathing on greenhouse roofs or evading aggressive pet cats, let alone treading on broken glass or eating leftover pizza. Wild nature has been twisted out of joint in Britain; except for the lonely saltmarshes of the north Norfolk coast, very little has not been reshaped by our fingerprints. But to understand the fox among us, we must first consider the world as it was before.

I AM FOLLOWING a wild wolf with a hind foot as wide as my hand. The paw print is written in the soft clay of a path flanked by wrinkled old trees that take circuitous routes to the sky, their branches bending under a red squirrel's leaps, their bark festooned with furry moss. Looking past their trunks, I see more trees, and yet more: shadows of green upon green, and woodpeckers laugh from within. There is not a sound nor sight save of natural things – just tree frogs purring at dusk, and pure, sweet, forest air. Where the canopy has been opened by a giant's fall, new saplings race to the light over the carpet of wild garlic flowers. Deer consume many such infant trees and are themselves taken by wolves; the leftover bones fall to foxes.

This is Białowieża, lowland Europe's closest match to a truly wild forest. It mantles the border of Poland and Belarus – a living palace of oak and hornbeam, and a mortuary of naturally dead trees that sport brilliantly coloured fungi, nourishing new life where they crumble. Old growth or primeval forests are relatively widespread in North America, although their fate is the subject of bitter battles between loggers and environmentalists. On the European Plain, we only have Białowieża, and its history has been uneven, from bison-hunting Russian Tsars to the Nazis who murdered Polish patriots under gently shadowing leaves.

Yet today, the rails laid by Germans to export timber in World War I are overgrown by wild pansies and chickweed. Disputes over logging in the buffer zone aside, the forest remains largely in control of itself, as it has been for most of the last 8,000 years. Human tragedies, triumphs and the entire Roman Empire have risen and fallen, and, all the while, Białowieża has quietly evolved along its own lines, its vast compendium of living things predating, competing, and joining into symbiotic relationships with each other. It is the last benchmark: a reference point that explains how European foxes are when the human touch is light.

That path where I found sign of the wolf's wandering is also a datasheet of information. The damp ground reveals more tracks – of wild boar, roe deer, red deer, bison – and here, not three metres from the wolf's trail, a fox footprint.

I could probably fit all four of its little paws into the track of its distant relation.

It is living here as a pure wild animal, sustaining itself on other life supported by the forest, and when it dies, its flesh will grow more trees.

What is a fox in such a place? I've caught glimpses on my camera traps: visually indistinguishable from foxes in

Białowieża.

gritty south London boroughs, they trot confidently down trails trampled by gargantuan bison bulls. They balance deftly on fallen tree trunks beside rivers that flood when they please. They lope across tracks at night, under crystalline stars untainted by light pollution.

There are questions that all wildlife biologists ask of their study species, and most of them would take a lifetime of research to answer. Where do they live, and how are they distributed within those areas? Do the regions with the highest activity have characteristics in common? What about their social interactions with their own kind – do they defend territories? When do they breed, and how far do the young disperse? In a forest with uncountable forms of life, with which of the plants, mammals, birds, reptiles, and invertebrates do they interact as predator, as prey, as disperser of seeds, as bringer of change through something as subtle as digging the mud to create a den?

Unravelling these mysteries requires deciding a research hypothesis, collecting data to test that hypothesis, and exploring those data with statistics or maps. I have no recourse for such ventures during my holidays in Białowieża. But I do spy some clues: fox scat, for example, the ubiquitous silent witness to their diet.

It is full of the bristling black hairs of wild boar.

Boar are as far beyond a fox's hunting prowess as is an elephant – fourteen times heavier, and armed with tusks and teeth. They are quite capable of fending off leopards. Yet here in a near-pristine forest, they comprise a startling portion of vulpine diets.

The link is the wolf. About 30 per cent of the fox's winter diet in Białowieża is carrion, specifically red deer and wild boar. Some may have starved, but many were killed by wolves and lynx. Despite the popular image of wolves as marauding Vikings, they inadvertently assist many other creatures. Ravens, martens and eagles are among those which feast on the wolf's leftovers.

The beauty of this forest is its completeness; nothing is wasted, and ecological relationships almost forgotten in Britain's radically altered landscapes shine clear.

Perhaps the message from Białowieża, then, is that the wild fox is an interactive creature. They are part of a living web that cycles energy from plants to herbivores to predators to scavengers, a natural dodgems in which species knock into each other. Sometimes that brings death to one and food to another; sometimes there is mutual gain. Wherever it is, whatever the backdrop, a fox can never be understood outside of its relationship with the rest of the natural world.

But what is a fox in the first place?

A FOX'S SKULL, like a wolf's footprint, fits neatly in my hand. In its way, it is a birth certificate, a genealogical signpost that reveals the fox's relationship to other mammals, living and extinct. Not for nothing did Victorian scientists spend so many hours measuring bones; patterns emerge from meticulous study, and the bewildering array of vertebrate life can be organised into logical groups. In recent decades, genetic analysis has refined our understanding.

So imagine an Ark with all the world's animals entering two by two: birds, reptiles, amphibians, and other, more exotic creatures. A fox would enter the deck reserved for mammals, for it has fur and suckles its young, along with a diaphragm and a neocortex. But a fox is clearly different from its mammalian relatives. It is certainly not a bat, giraffe or horse. To divide mammals logically, Noah must look at their jaws. Teeth are not only clues to diet, but also indicate who is related to whom.

Foxes have 42 teeth inside their narrow muzzles, among them the trademark of the order Carnivora. Their last upper pre-molar and first lower molar are called *carnassials,* modified with a tall shearing edge for slicing flesh. All 300 or so Carnivorans possess carnassials. Your pet cat has them – in fact they are exceptionally well developed in felids, which are the most meat-specialised of the entire order. Pandas, on the other hand, have flattened carnassials because they are largely vegetarian.

What about foxes? Their carnassials are specialised for nothing. Berries, mice, insect larvae – they can eat it all. That adaptability is the main reason why the wild dog family has been such a runaway evolutionary success.

Foxes are of the canine kind. Many fox-watchers, including myself, enjoy the cat-like grace of the species: the delicate pounce and careful footsteps. But beneath that rich orange fur, foxes are undeniably a canid – that is, a member of Canidae, the dog family. Turning the skull over, the base of the ear sockets is fused into a bony casing called a tympanic bulla, protecting the fragile inner-ear bones. In canids it is uniquely divided by a septum: a thin, bony wall. It is believed that this extra echo-chamber enhances their ability to hear low frequencies.

Five toes with claws that are not fully retractile, a long muzzle filled with delicate turbinal bones that magnify smells, the nuchal ligament that strengthens their necks to allow them to run for extended periods with nose to the ground – the physical hallmarks are unambiguous. The fox is a dog.

Or very nearly.

THERE WAS ONCE another forest, more remote to our lives and fainter than Białowieża's leathery trunks: an ancient jungle that no human saw, in a global climate hotter than we have ever known. Fifty million years ago, the so-called Eocene hothouse featured Earth at its balmiest, basking in temperatures averaging 14 °C (57 °F) higher than present. Ice vanished from even the poles, and lush forests extended across the globe. Palm trees grew in what is now the London basin; extinct primates foraged under them, and the peculiar horse-like ungulate *Hyracotherium* browsed in their shadows – but there were no foxes. Canids had not yet left their family cradle.

That birthplace was North America; perhaps dogs are the United States' most successful export. Canids have spent the majority of their evolutionary history restricted to this one continent, and the distant ancestors of foxes are dated to the Eocene jungles of Texas. *Prohesperocyon* – the first known canid – was a small, omnivorous creature in a forest of astonishing giants, a natural neighbour to the likes of the fearsome Nimravidae, sabre-toothed carnivores resembling great cats. As Earth gradually cooled and dried and the dense jungles were replaced by extensive grasslands, the elongated legs and long strides of the canid body shape clearly provided an advantage in long distance chases. Canidae thrived.

Most of the fox's extended family is extinct, and known to us only through palaeontology, but the glimpses defy imagination. Among them is the subfamily Borophaginae, containing species that might frighten Cerberus – they were the bone-cracking dogs, with massive teeth capable of extracting marrow from giant carcasses. *Epicyon haydeni,* the largest of all, is estimated to have weighed four times as much as an average grey wolf, and a pack on the hunt must have been a formidable sight. Yet the dentition of many fossil canids, not least the little *Leptocyon* – the direct ancestor of the fox – strongly suggests a varied, plant-inclusive diet.

Plate tectonics and the reduction in sea levels during ice ages eventually connected North America to other continents, and its wildlife migrated into South America and Eurasia. By seven million years ago, at least one canid was present in Spain, and *Vulpes riffautae* – the earliest known fox outside North America – lived in what is now the N'Djamena desert in northwest Chad. Fast forward another few million years, and *Vulpes galaticus* is part of the Turkish fauna. *Vulpes vulpes,* the red fox – the only species that we have in modern Britain – is recorded first in Hungary, perhaps 3.4 million years ago, when humans in Africa were just beginning to use stone tools. Recent genetic analysis has provided further hints; it appears that all living red foxes are descended from individuals who lived in the ancient Middle East. From there, they spread across the entire northern hemisphere.

Canids travel. Their long legs and unfussy diets enable them to colonise new habitats with ease. Nature is not a fixed condition, and Britain has experienced many waves of colonisation and extinction over its geological history. But from the perspective of a wild animal, one of the most significant qualities of our island is that it lies rather far to the north. Much as we complain about the weather, it is remarkably mild for a country on the same latitude as Moscow. Take away the Gulf Stream, and it would be time to buy some serious winter clothing. Add all the geographical and solar phenomena which regularly cause the world to have ice ages, and the Thames Valley becomes cold, hard tundra.

I have tried to breathe normally in temperatures of -35°C (-31°F), on a day in Alberta when humans seemed disinclined to be outside. Gulping such air is like swallowing swords; your lungs baulk at the freezing blast, yet it is nothing compared to the frigid temperatures reached in the wind-chill from a continent-wide ice sheet several kilometres thick. The Pleistocene Epoch played a long

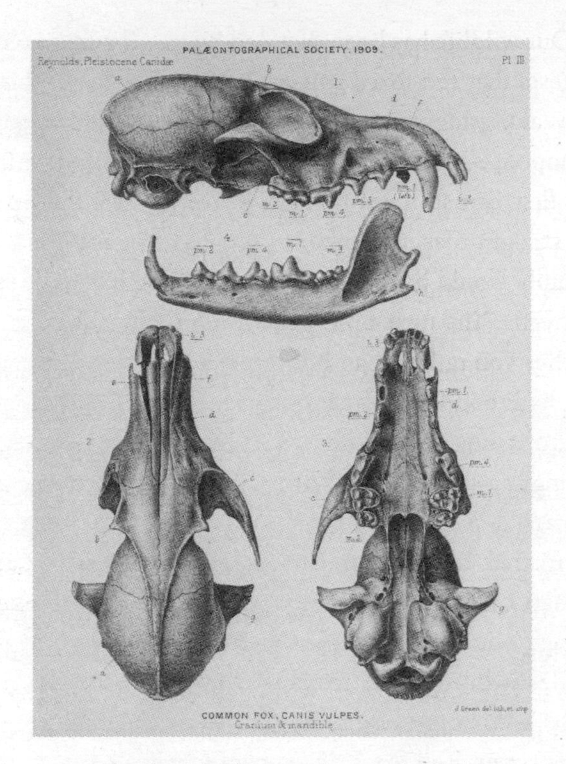

A study of red fox bones. This one lived during the Pleistocene in what is now Kent.

game of catch-and-release with Britain, repeatedly coating the northern half of the country with dense ice and then releasing it in warm interludes called interglacials. We live in an interglacial now called the Holocene. It has lasted nearly 12,000 years, but it probably won't continue forever.

Our wildlife has been dictated by ice. The fossil record suggests that red foxes appear in the interglacials, historically alongside a wealth of other creatures that would monopolise attention if glimpsed on safari in Africa. Our first British foxes perhaps scavenged on the carcass of a straight-tusked elephant predated by cave lions, and certainly would have heard the whooping laughs of spotted hyena. The next time you wonder why a fox sits and watches you rather than bolting in panic, remember they have had to judge the risk from very dangerous predators for thousands of millennia, and their evolved strategy is waiting at a safe distance with access to a known safe spot, such as a den or – these days – a gap in a fence. If they had run further than required from each Pleistocene sabre-toothed cat, European jaguar and cave bear, the energy wastage would have crippled them.

Meanwhile, Arctic foxes – along with woolly mammoths, wild horses and reindeer – were present during the colder times. Red foxes disappeared from Britain entirely, surviving in the relatively mild refugia in Spain and the Balkans. Sometimes, while walking in my native Surrey Hills, I try to imagine what those glacial millennia must have been like. The glaciers never extended this far south, but the bite from the wind must have been excruciatingly bitter, and the landscape would have been an austere mixture of bare rock and frozen snow. Fed by the ice sheet were huge rivers of cyan-blue glacial meltwater.

And foxes must have drunk from them. But were they the Arctic or red species?

Over in Somerset, one cave dated to 12,000 years ago contains fossils of both, but red foxes have always returned north and displaced their smaller Arctic cousins in times of mild climate, and they continue to do so on the modern thawing-line of Sweden. In any case, while lions and hyenas did not migrate back to Britain after the ice retreated, foxes rapidly did, even as tundra budded with crowberry bushes and mugwort, and finally grew trees once more.

So the fox trotting across a Clapham street is directly descended from individuals that encountered species wondrous beyond our most outlandish fairy stories, survived extremes of climate that we have never known, and crossed land bridges long lost beneath the sea. Human culture is such a late entry into the story of the fox that it would seem disingenuous to mention it – except, of course, we have a strong bias towards it.

No one will ever know where the first *Homo sapiens* laid eyes upon a living fox, or how the two species perceived each other. As pre-history continues, our fossils and theirs begin to overlap in palaeontological sites, a silent testimony to forest meetings that have passed into the veil of unwritten time. But 16,000 years ago, when Palaeolithic painters were drawing steppe bison in the Spanish cave of Altamira, a woman of unknown name died in what is now Jordan, in a site called 'Uyun al-Hammam. Her body was laid among flint and ground stone, and a red fox was carefully placed beside her ribs, resting with her for eternity on a bed of ochre.

We cannot perceive the meaning. Was this a pet, or an animal kept for its ceremonial significance? The care in the joint burial is believed to suggest some emotional link between human and fox, beyond that shown to wildlife perceived as food or clothing. It has been speculated that these pre-Natufian people coexisted with foxes that were at least half domesticated. Perhaps they scavenged rubbish on the edge of camps, along with the earliest dogs. Perhaps the behaviour so often complained about in London is more ancient than we think.

In any case, it is clear that foxes held a strong cultural significance for the later peoples of the Levant. They are commonly found in human graves in Kfar Hahoresh (modern Israel), dated to around 8,600 years ago, while stone carvings of foxes with thick brushes adorn the pillars of Göbekli Tepe in Turkey, believed to be the world's oldest temple. In Mesolithic Britain, humans who hunted deer by the shore of extinct Lake Flixton – in the North Yorkshire archaeological site of Star Carr – must have been aware of their small red neighbours. Bones from two foxes have been found at this ancient settlement, along

A nine-tailed kitsune in nineteenth-century Japanese art.

with those of Britain's first known domestic dogs, but there is no indication of what role, if any, canids played in their culture.

Later, as humanity discovered the joy of story-telling, foxes joined the cast. The oral literature of native Americans occasionally opts for a fox as a trickster, albeit a potentially handy one; according to one Apache legend, it was Fox who stole fire from the fireflies and introduced it to Earth. It is across the Pacific in Japan, however, that fox folklore reaches its most astounding heights. Kitsune – the revered fox of Japanese myth, poetry and traditional belief – has existed in human thoughts for many centuries. It even makes an appearance in what may be the world's oldest novel: the eleventh-century epic *The Tale of Genji*, where a human character debates whether the figure by a tree is a woman or a shapeshifting vulpine. Kitsune delight, deceive and confuse in countless other legends; while the theme of pretending to be an attractive woman is frequent, other tales relive how they mislead travellers by lighting ghost fires at night, assume the form of cedar trees, or even become the guardian angels of samurai. Today, anime writers continue the kitsune tradition.

BACK IN EUROPE, by Roman times the uneasy relationship between foxes and agriculture had woven itself into religious rituals – in the festival of Cerealia, for example, live foxes were released into the Circus Maximus with burning torches tied to their tails. Seven hundred years later, Aesop's tales also provide a nod to fox interactions with farmers, and – to a lesser extent – with their neighbouring wildlife. My favourite Aesop fable features a wolf taking a fox to court for theft; given the vast quantity of wolf-killed carrion that real foxes consume, it seems vaguely reasonable.

Old English literature picks up similar themes. *The Fox and the Wolf*, a rhyming poem from the thirteenth century, stars a fox who helps himself to some chickens and then tricks a wolf into taking the blame:

A fox went out of the wood
Hungered so that to him was woe
He ne was never in no way
Hungered before half so greatly.
He ne held neither way nor street
For to him (it) was loathsome men to meet
To him (it) were more pleasing meet one hen
Than half a hundred women.
He went quickly all the way
Until he saw a wall.
Within the wall was a house.
The fox was thither very eager (to go)
For he intended his hunger quench
Either with food or with drink

And so it continues, with the hungry fox trapping himself in a well before deceiving a wolf named Sigrim into taking his place. Ironically, this poem was written about the same time that the wolf's howl was finally falling silent in southern Britain.

Did the fox notice the disappearance of its distant relative? Perhaps, unconsciously. As shown in Białowieża and elsewhere, the wolf was a provider as well as rival, a powerful force in the wildwood whose absence has changed these islands as much as a spoke missing from a wheel. Some species have sharply increased, and others have probably declined.

Yet civilisation has done more than simply rip out culturally troublesome natives while boosting deer and grouse for hunting. We have released millions upon millions of non-native animals into the countryside: rabbits from Spain, fallow deer from Persia, sheep from Mesopotamia, hens from south-east Asia, cats from Africa. Our trading ships accidentally added black rats from India and house mice from the Middle East, while American grey squirrels, Japanese sika deer and even Australian red-necked wallabies joined our countryside from zoos. We have persuaded ourselves that the six million sheep of Scotland are part of the 'natural' scene, but the Highland ecosystem evolved with none. Even the Scottish red deer population of 300,000 is far higher than in the time of the wolf. These changing grazing pressures affect the rodents and berries that foxes eat, and near-total deforestation has altered their territory sizes and feeding habits.

In a flash of geological time, we have rewritten the

fox's wildwood, in ways both graphic and subtle. We have added, taken away, replanted and concreted.

And the fox that once played its natural dodgems with the rest of the natural web will inevitably interact with the components of the new Britain that we have designed without ecological aforethought.

The fox is not an intruder into our world.

We have simply laid our modern ambitions over the landscape it already knew.

3

Where Do Foxes Live?

OPEN-TOP TOURIST BUSES and impatient black taxis battle for territory in the concrete canyons of central London; beside the gridlock, cyclists squeeze past wary pedestrians, and silent women push today's *Metro* into the hands of freshly arrived train passengers. The city's heart is within the embrace of the two highest towers of British justice: the Royal Courts with its soaring gothic spires and vaulted archways, and the Old Bailey, centuries-old theatre of the grimmest human drama. Perhaps it is no surprise that such a place tries to judge foxes too.

Humanity floods the senses. It's noisy, so noisy, with cars, and drills, and cries of 'Can I interest you in a . . .'

Salesmen offering free organic yoghurt samples, those you can escape; not the smell of vehicle exhaust, however, nor the tourists agog at military statues that screen out so much of the sky. It is musty yet grand, the mood here: intimidating, disconnecting and mesmerising. It is the bones of something; British history, perhaps, stacked so high over press crews hoping to witness more of it, while a tiny old man tries to photobomb them – his Staffordshire bull terrier is wearing a jacket emblazoned with anti-nuclear slogans.

For British people, these streets are a hook from which we dangle and debate our civilisation. For British foxes, this is a land of nothing.

Truly, nothing at all. Not a blade of grass, not a mouse, and hardly a bird in the sky. The ancient wildwood has been utterly extinguished.

At least, all logic would say so.

Yet there was a fox in this very place, not many hours past – a single scat has been deposited on a sprawling gum-spotted pavement between a bus shelter plastered with anti-police propaganda and the unsmiling security fence of the Royal Courts of Justice; a homeless man begs for coins from a populace oblivious to both wildlife and him. Over towards City Thameslink, where wet concrete was recently laid to tidy some aberration, a fox footprint is written literally into London's frame.

That foxes thrive in leafy suburbs, wooded gardens, and even fields radically transformed by intensive agriculture is not news. But the Strand is a frontier beyond most living things. Faded carvings of red squirrels brighten one business's wooden sign, while the tavern's name leaves no doubt that cockfighting once brought brawls and gambling within a street-sweeper's walk of the Inner Temple. But there is not much non-human life today, save for the pigeons where tourists break the law and feed them, and a gull or two chortling from the spires.

And a fox, somewhere.

When people exclaim that foxes are everywhere, they are both correct and imprecise. The Mammal Society's National Mammal Atlas shows fox records in nearly every British grid square, from Cornwall to Sutherland, the Cambridgeshire fens to the Western Highlands. They are absent from the remotest islands, but mainland Britain is unquestionably the domain of the fox.

Not to an even degree, however. The weakness of a simple presence-absence map is that it gives the impression that all landscapes are equal. In reality, of course, some places have far higher fox densities than others. No one considers it surprising that humans are clustered tightly in cities, with only a smattering of houses in moorland. In one sense, the fox population also has its high-rise flats and hamlets. All land is not alike.

'DESERT FOX!'

My guide, a khaki-clad middle-aged Indian of military bearing and Sherlockian skill in tracking all creatures

Crossing India's Rann of Kutch, part of the Thar Desert.

wild, spins the steering wheel of our jeep. Dust spurts, the vehicle's suspension lurches, and behind us lie treadmarks in the white, white dust. Ahead, no doubt, there is a fox; but mostly there is nothing, as only the desert knows it. Vast flat horizon and vast, vast dusty sky: a land crossed by Rabari tribals and their cattle, but immune to the modern world. I am in the Rann of Kutch in India's Thar Desert, rattling across the dry ancient bed of the Arabian Sea. I have travelled to many remote places, but this is a landscape apart: seasonally cracked in fiery heat, swamped by monsoons, bleached by salt, and blurred by mirages – stark, wild, beautiful and brutal.

The jeep has stopped.

A fox looks up at me.

It is sitting in a scrubby thicket of what the local people call toothbrush bushes, amber eyes so clear and sharp. It is a red fox, *Vulpes vulpes*; just like those in London, although its fur is straw-coloured, as if irradiated by the Gujarati sun. It is a curiously sobering thing to observe a fox in an overpoweringly enormous landscape – a theatre refined by torrid heat until it retains only the core essentials of grit and sky. They, too, are raw and unhumanised, and their basic needs cleanly defined.

What is actually needed for survival? We ask that question of ourselves in *Robinson Crusoe* and its modern spin-offs, but applying it to wildlife may remove the confusion over seeing a fox in the very heart of the metropolis. A hypothetically shipwrecked fox would probably thrive, for its needs are very simple: some shelter to evade weather and enemies, and about 120 kilocalories per kilogram of bodyweight per day. That equates to about nine voles or one rat daily – or one double cheeseburger with fries. Even the bleakest of our cities offer sustenance on this scale to a scavenger-hunter.

The cracked dust of the Gujarati desert does support some hardy plants, which in turn feed herbivores. The desert fox may seek exotic-sounding rodents such as the golden bush rat, the jird, and the Indian gerbil; insects, and the carrion left behind when wolves or jackals kill chinkara gazelle, are also possibilities. Those little thickets of toothbrush bushes – known as *bets* – offer shelter from the murderous May sun and stay above the waterline during the monsoon floods. Nothing more is required. However improbable it may feel to a human figure dwarfed by a blood-red sunrise, watching wild asses gallop across bone-dry salt flats, this land is perfectly suitable fox country.

On the other hand, so is the ancient forest of Białowieża in Poland, where bank voles scurry past gigantic fungi and wolves inadvertently provide a regular feast of wild boar carrion. So are the gloriously wild prairies of southern Canada, where a bewildering array of rodents whistle from meadows painted glittering silver in springtime ice storms. And so most certainly are suburban British gardens, where they may have their weekly calorie requirement handed to them on a dogfood dish every single night.

The *abundance* of potential food in each of these habitats is different, however. There is no 'normal' or 'correct' fox population. Each area is unique. Even the subtlest local changes can trickle upwards – in the harsh mountains of northern British Columbia, for example, areas dominated by lichens are avoided by foxes in favour of those where goat willows are found. In Belarus, forests growing upon clay soils support more prey than those on sandy deposits, and have higher fox densities. How many journalists musing over British fox numbers have thought to take samples of the local soil type?

Obviously, the more food available, the more foxes that the area can potentially support. As a general principle – and notwithstanding countertrends driven by disease and the impact of natural competitors like badgers and coyotes – foxes are distributed unevenly across their huge natural range because food itself is uneven. By that yardstick, the Strand may be even harsher than the Thar Desert; yet both have their foxes. So do the Himalayas, the sub-Arctic, the rainy Spanish mountains and Edgware tube station.

At this point, it is worth taking pause. Think of the world's most famous animals: tigers, elephants, koalas. How many exist in a range of habitats even close to the diversity of the fox's natural homes? Range expansion is one of the fox's rewards for being unspecialised.

Improved odds of beating extinction are another. Replacing wildwood with cold London stone devastated many of our native species, but the fox has survived – and often thrived – during all our changes to the British landscape. A clue as to why comes from the enchanting knife-edged mountains of Sichuan in China; unlike the giant pandas that also wander this landscape, Sichuan's foxes do not risk starvation when a single food source fails. The panda, famously, is a specialist consumer of bamboo. Should this plant flower and die, as stands do on a regular basis, the panda must move to a new area or perish. Not the fox with its catholic tastes; if, say, its stereotypical British prey of field voles runs short, it will simply switch to pouncing on wood mice or rabbits instead.

Nor are they specialised to a specific habitat. Otters can be wiped out from an entire district by river pollution.

A fox population, in contrast, covers so many habitats that even if it faces an environmental disaster in farmland, it will persist in the neighbouring wood, and soon recolonise.

Wherever it lives, a fox learns an acute cartographical knowledge of its local landscape and explores it at a purposeful-seeming trot. In the Swiss Jura, foxes travel about 4 to 12 km (2.4 to 7.4 miles) daily; interestingly, their kin in a residential district in Toronto, Canada, have wider extremes, varying from 2 to 20 km (1.2 to 12.4 miles). Urban Canadian foxes are provided with far fewer deliberate handouts than their British counterparts, however, so source a large percentage of their meals directly from the land.

While foxes have allegedly been clocked at 50 kph (31 mph) in short bursts, their typical pace is far slower, and punctured by rest periods in which the fox will doze under a hedgerow or in a quiet urban corner. The Swiss foxes averaged a speed of about twelve metres per minute, although one individual, who was a transient – a fox without territory – moved considerably faster. While all this may seem like a considerable exercise regime, it is far below the 26 km (16 miles) averaged by male wolves per day. Individuals of both species that are dispersing from their parents into a new territory can wander much further.

One persistent piece of fox folklore is that they are

nocturnal – that is, active by night only. Sometimes, this myth slips into the medical department via warnings that a fox enjoying the sunshine must be ill. In fact, it is no cause for alarm. Foxes do pursue a nocturnal existence in regions where they are heavily persecuted, and, as is the case for many human-caused aberrations to the natural world, we have grown accustomed to this atypical state of affairs and convinced ourselves that it is normal. Left to their own devices, foxes will adapt their activity patterns around their social and food-gathering needs. In the world's great wildernesses, from the Thar Desert to the boggy forests of Ontario, foxes are easily found abroad during daylight hours.

Foxes are often active in daylight where they are undisturbed.

In Britain, field voles tend to be diurnal – day active – if the temperature drops below freezing, and foxes, and indeed barn owls, naturally follow suit. Needless to say, if they find a person who regularly feeds them pork sausages in daylight, they will adapt their activity around that food source instead. I have also known several low-ranking foxes that opted for daytime travel to avoid confrontation with dominant individuals.

Radio-tracking has shown that the daily wanderings of a territory-owning fox fall into two distinct types. The first is a circumnavigation of its entire territory, and the second – and more common – is of visits to different parts of their range each night. Varying their journeys gives them the optimum chance of exploiting food resources; if they were to concentrate on the same field month upon month, it would eventually run dry of voles while the untapped pasture half a mile away is awash with them. It is worth adding that the enormous bounty provided by people who feed garden foxes has added a third trend: foxes who travel little and appear in sizeable numbers in specific sites every day.

Under more natural conditions, foxes tend to cross the landscape in a large-scale zigzag pattern. They are often religiously loyal to specific routes, wearing narrow paths into grass through repeated trampling. In the wilderness, they climb onto fallen tree trunks and walk down their full length as a kind of elevated track; in Britain, they occasionally exploit railway lines instead. Last year, I was shown some startling footage of a fox in Wales trotting briskly down a train rail hardly wider than a human hand, balancing like an expert on a tightrope.

Railways and foxes often occupy the same sentence. While human commuters frequently feel that our rail network is more of a hindrance than a help to travelling, it is commonly stated that our vulpine neighbours are transported by them. Not as passengers – although there are several credible accounts of urban foxes jumping on board public transport – but rather as walkers along the banks. Even at those moments when the Gatwick Express thunders past the East Croydon congestion at 100 mph, and on just the other side of Network Rail's perimeter fence, millions of people shop, argue and check their phones, the embankments themselves remain one of the least disturbed environments in the city. It is often said that foxes first immigrated to London in the 1930s, the pioneers moving down railway lines into this new brave world full of human creatures.

But when talking of the arrival of fox in city, it is as well to remember that city has also travelled extensively into the traditional land of the fox. London has bloated massively over the last two centuries, and despite the best efforts of greenbelt campaigners, continues to do so. Many of today's 'urban' foxes may be descended from 'rural' foxes whose habitat was suddenly turned into housing estates. Incidentally, records of foxes near towns in Finland date back to the medieval era, and their distinctive barks were heard in Tokyo in the ninth century. There is even some suggestion that foxes scavenged on abandoned scraps from humans as long ago as the Palaeolithic – the Old Stone Age.

Regardless, considerable research has taken place in recent years to establish the impact of railway lines on fox densities and movement patterns. But evidence that dispersing foxes, and indeed territory-owners on the hunt, are funnelled by the railways is also remarkably scant; radio-collared foxes have shown little preference for the train lines.

Taking the wider view, why would they? Humanity has proved tragically skilful in fragmenting the habitat of hedgehogs, toads and dormice, but foxes are much more capable travellers. They can and do cross roads, car parks and fences. Even the natural world's topography has little impact; genetic sampling from Croatia has shown that they migrate freely across rivers and small mountains.

When not travelling or feeding, foxes require a suitable place to rest. This may be anywhere within their home range, even close to the territorial border. Foxes have more than one den, including sites that may only be used temporarily. Researchers in Polish farmland found that earths tended to be dug on steep south-facing slopes, with western exposure avoided, possibly due to the prevailing westerly winds. In suburban Britain, foxes often rest on greenhouse roofs or sunbathe in quiet alleys. I've found one stretched out contentedly on warm plastic in a narrow gap between a wall and a garage, peaceful and safe, despite being within metres of a major supermarket car park used by hundreds of humans each day.

BUT THE STRAND is crossed by thousands upon tens of thousands of people. Even at midnight, it is alive: lights on the arching stonework, music thumping from cars as they choke in bus-filled traffic jams. I've come back here because I want to better understand the miracle of foxes in a desert of towering grey rock. Friday night has spilled people upon the streets, shouting, selfie-taking, watching buskers batter their drums. More of them sprawl between the paws of the giant lions of Trafalgar, strangely drawn to the cold stone models of wild animals. Night itself seems defeated by the battle squadron of lights jumping upon these grandest of buildings, reflecting on the river, luring punters into shops. So it continues on the journey southwards: shuttered shops, drunken youths, urban cries and urban dreams.

There is a fox.

A male, all long limbs and thick brush, sits on a patch of grass under one of Brixton's tower blocks, half illuminated by streetlight. He turns his head towards the car as we pass, watching, just like his kin in the silent and utterly wild Thar Desert.

Adaptability.

That, in essence, is the fox's gift.

4

What Does the Fox Look Like?

> The world is full of magic things, patiently waiting for our senses to grow sharper.
>
> W. B. YEATS

IN A LONELY ARABLE FIELD in eastern Surrey, the North Downs Way National Trail and the Prime Meridian collide in a crossroads perfectly aligned to the points of the compass. To the south lies the ceaseless rumble of the M25 and its luxurious trim of rolling green countryside. Northwards rises the chalky ridge of the Downs – due north, in fact, and I know this from my map and the position of the sun. But if I were a fox, my navigation might be written into my physical senses.

Slightly offset from true north, the Earth's magnetic north pole drifts each year due to the behaviour of molten iron in the planet's outer core. The invisible magnetic field that envelops our planet protects us from harmful cosmic rays while also playing host to the geomagnetic storms that produce spectacular auroras: the northern and southern lights. It has had a profound impact on human history, because compasses – those aids to explorers, traders and armies for millennia – only work because their magnets swing to the poles, where the field's inclination is vertical.

The geomagnetic field and its poles also have very real significance for animal behaviour. Birds may navigate by it, rats become more restless during magnetic storms, and some researchers have argued that resting cows tend to align themselves pointing poleward, except under power cables which locally disrupt the field.

And foxes, perhaps, hunt by it.

Imagine that you need to catch a rodent in dense cover. The rodent, naturally, does not want to be caught, and is equipped with formidable defensive senses of its own. It may also have awkward behaviours; bank voles, for example, reduce their activity upon detecting fox scent, for the rustles of their feet on vegetation are, perhaps aptly, their Achilles heel. But hearing may not be the fox's only means of pinpointing its target.

A fox that is stealthily approaching its intended meal will be most successful if it orientates itself either within about 20 degrees of the magnetic north, or due south, at least according to one recent study. Leaps from other directions usually fail to pin the prey. If foxes are indeed capable of magnetoreception, the mechanism by which they perceive the direction of the magnetic poles is unclear. The authors of this study speculate that foxes perceive the geometric field as an area of light or shade in their vision – in fact, even in people, laboratory tests show that the field impacts light perception.

Foxes may use the directional information from the magnetic field together with auditory input from the vole's rustling to move to a fixed distance from the target, allowing a precise leap. If so, foxes are the first species known to use the magnetic field as a measure of distance.

THE PHYSICAL FOX – the frame that supports this curious, intelligent, beguilingly strange canid lifeform – is complex. That frame is aesthetically pleasing to many human observers, but our perception of beauty is irrelevant to the creature itself. It is the demands of survival that have whittled the fox's senses, size, bone structure, brush shape and teeth.

What has emerged is almost the perfect formula: a carnivoran that is omnivorous – a generalist – yet which still carries a specialist's trump card. As we have already seen, the fox can survive almost anywhere, and on a mind-bogglingly diverse array of food; but for all its catholic tastes, it has never lost its finesse in hunting rodents. Much of what we admire about the fox is a direct adaptation to the challenge of catching such small, swift prey. In a very real way, foxes are built around mice.

It is part of their bones – especially in North American foxes, which have limbs considerably lighter than expected for a canid of their size. A light frame with a small stomach can be launched with ease at a rodent-sized target. The huge tail aids balance.

Foxes are guided by their hearing, which is sensitive to a degree that human imaginations might leave short. Every winter I observe foxes hunting field voles in frosty meadows, weaving slowly through quiet tussocky grasses in a gentle amble so different from a travelling fox's precise, purposeful trot. One morning, I was watching a handsome russet fox in a sloping field, when he veered sharply to his left, tilted his head – raising one ear canal above the other to better pinpoint a sound's position – walked about 6 metres (20 feet), and pounced on a rodent.

A fox can hear a much wider range of frequencies than us; their upper range is similar to a dog's, while the lower range is comparable to that of a cat. Their eyesight is much weaker than ours, but photographing a fox at night with a flash uncovers one of its ocular secrets. Unless you are very careful or fortunate, the resulting image will feature a coloured washout of the animal's eyes. The culprit is

A fox sees, smells and hears the world very differently to us.

the *tapetum lucidum*, a remarkable layer of tissue directly behind the retina. Present in many creatures that are active in low light levels, from fish to tigers, it reflects visible light back through the retina, effectively brightening the world for its owner.

Domestic dogs possess a *tapetum lucidum* too, but in another aspect their eyes are very different from those of a fox. In bright light, a dog's pupils contract to a round shape, which is not surprisingly also the case for the ancestral grey wolf. But a fox's pupils contract vertically, like those of a small cat. Is this an advantage? Absolutely. Canid eye lenses contain concentric zones of different focal lengths, and a vertical pupil can exploit all these zones even when at its narrowest in bright light. This improves the focusing of long and short wavelengths of light, reducing or eliminating chromatic blur/haze in bright conditions. In short, foxes are multifocal.

More differences from dogs are under the skin. Fox skulls can appear superficially similar to those of dogs – I recently had to confirm the identity of one unusually large fox skeleton by examining whether the pre-orbital processes in its muzzle were concave or convex – but it is always worth considering the sagittal ridge. This runs vertically across the top of the braincase, and in case you are now rubbing the top of your head, please be reassured that you do not possess one. It is found in some great apes, however, as well as many Carnivora species, and even some dinosaurs. Attached to the sagittal ridge are the temporalis muscles, which are used for biting. The bigger the ridge, the more powerful the snap of those jaws.

Many thousands of years ago, foxes shared North America with a huge canid called the dire wolf. We have

Skulls of a dire wolf (left) and a red fox.

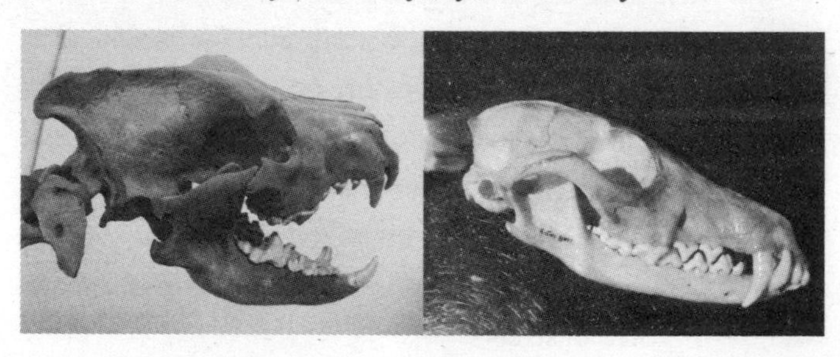

learned of this species Sherlockian-style, piecing its life together from minute details of the bones it has left behind, mostly in the macabre La Brea Tar Pits of California. This lethal but fossil-rich site has gifted museums a grand surplus of dire wolf skeletons. One is now on display at the Natural History Museum in London, and I recommend pausing beside it should you be visiting South Kensington, for the sagittal ridge jutting out from its skull cuts an eye-wateringly impressive flange. A mighty hunter that pursued bison and elephants, or a mighty scavenger that crunched the bones of carcasses left behind by giant cats; we will never know. Regardless, its jaws packed incredible power.

Foxes don't hunt mammoths. Their sagittal ridge is remarkably small; in fact, their bite is weaker than that of a dog of the same weight, a fact exploited by many huntsmen over the centuries. Some dog breeds bred to kill foxes, such as Jack Russells, are noticeably smaller than their targets.

But just how big is a fox? It is a complicated question because they play an optical illusion on the human brain: they appear larger when they are further away. Some of us suspect the same phenomenon occurs with pet cats, which suddenly become 'black panthers' and 'Beasts of Bodmin' when sighted across a foggy meadow. Regardless, despite media speculation, there are no foxes the size of Labradors wandering British streets.

Measured objectively, a fox often weighs less than a pet cat: the average weight of male foxes in England is 6.7 kg (14 lb), with vixens about a kilogram (2.2 lb) lighter. Scottish foxes are somewhat bigger, with males tipping the scales at an average of 7.3 kg (16 lb). Of course, this is not due to over-indulgence in haggis, but rather an example of Bergmann's rule – the ecological principle that in a species with a broad geographic range, individuals in colder climates are likely to be larger.

Head and body length averages 62–65 cm (24.4–25.5 in.) in English foxes; for comparison, a typical domestic cat is around 46 cm (18 in.). The thickly furred brush is equal to about 70 per cent of the fox's body length.

ASKED TO DESCRIBE a fox, a person will almost certainly refer to its colour. The standard fox colouration is of course rich orange on its upper parts, with a white chin, throat and belly, and legs darkening to near-black. The brush usually but not always has a white tip. The hindquarters are usually more muted than the front.

In Britain, the vast majority of foxes exhibit only minor variations on this theme. There is a spectrum of orange hues, ranging from dark brown-grey to vivid carrot, and a few are noticeably greyer. There is a general perception in old literature that the foxes of northern Scotland are less russet. It is true, as discussed above, that Scottish foxes are on average somewhat larger than their lowland cousins.

There are no noticeable trends between vixens and dogfoxes. The old myth that only the males have white tips on their brushes is easily disproved. In fact, even practised observers can find it difficult to distinguish between the sexes at a distance, unless it is a season where the testes or teats are obvious. The most reliable clue is not a fox's colour but the shape of its skull; vixens tend to have narrow V-shaped features while a dogfox's broader cheekbones result in what has been compared to a W-shape. However, this is most marked in mature males; yearling dogfoxes can superficially resemble vixens.

Very rarely, one of the spectacular colour morphs caused by recessive genes surfaces and lands a fox on the local or even national news. *Melanism* is well known in wild cats such as leopards, where the dark form is commonly (and incorrectly) called the black panther. Genuinely melanistic foxes are highly unusual in the UK, but there are a few records from London and the Home Counties. They are much more common in North America, along with a part-melanistic variant known as the cross fox – a strikingly handsome animal with orange upperparts and a black face and belly. Another colour morph a few steps down from full melanism is having an orange back and dark chest; this form does occasionally

A black fox painted by the nineteenth-century American naturalist John Audubon.

appear in Britain, and I have twice seen it in my part of the Surrey Hills.

At the other extreme, foxes can be dazzlingly white. If you should spot a snowy fox trotting across your British lawn, this does not necessarily mean that an Arctic fox has escaped from a nearby zoo. It is probably a wild red fox that is demonstrating *leucism* – a sharp reduction in pigment levels caused by the failure of the melanocytes to migrate during cell development. The result is a mesmerising animal that is almost entirely furred in white; it may still show black 'moustache' stripes, or even have a noticeably whiter-than-white tip on the end of its very pale brush. Its eyes are normal hazel-brown.

The press sometimes claim that leucistic foxes are one-in-a-million, although absolute statistics are nearly impossible to calculate. They are certainly uncommon. It was therefore with high surprise that I received a tip in early 2016 that one had been sighted in my nearest town. Naturally, in twisted vulpine humour, this exquisite fox was not dwelling in a pathless ancient forest or even the town park – it was ensconced in a junk-filled alley between a skateboard club and a major supermarket car park.

That did not stop me racing straight round with a trail camera.

'You will need to think about thieves,' the caretaker of the nearby office block cheerfully informed me. It should be said that as an ecologist who operates camera traps professionally, I am acutely aware of the offence described with dry subtlety in British law as 'theft by finding'. To date, I have had eleven trail cameras stolen. These specialised cameras work, of course, by being placed on a convenient tree or fence and are triggered to take short movie clips by whatever passes by. From a scientific perspective, they are marvellous things that have revolutionised mammal research – provided no 'finders' with itchy fingers happen across them first, of course.

Standing in that alleyway with the clunks of car doors and rattle of supermarket trolleys in my ears, I felt a long

A rare glimpse of a leucistic fox.

way from my old trail camera research site in the wilderness of the Canadian prairie. Yet even the high crime vibe could not deplete the mystique of a white fox.

Many ecologists discover that they spend more time buying equipment in hardware shops than from nature-focused companies. Before that week was out, my poor trail camera was preserved with a steel security case, a saw-proof bronze chain whose rings were almost as wide as my hand, and a padlock worthy of Broadmoor. I tried not to imagine how long the camera would stay there if I should happen to lose the key.

But it worked. The reward was a very rare close-up of a startlingly beautiful wraith of a fox: creamy white with just a hint of ginger about the paws, clear dark eyes, and black moustache-marks on its muzzle. He was a yearling male, one of so many in that town; his family trotted past the camera too. All were regular orange hues and displayed no interest in their relative's unusual pelage.

Genuine albinos – individuals with pink eyes, with no pigment in their fur at all – are extremely unusual among wild foxes, although there are records from Spain and Canada. Part-leucistic foxes occasionally occur and may show white 'socks' on the lower part of their hind legs.

Whatever their colour, foxes fit the landscape in which they dwell. Their physical experience of the world is very different to ours, but it sustains them admirably across the full measure of the world's diverse environments.

5

Fox Family Matters

THERE IS A FOX CUB ON MY HEAD.

It wasn't how I intended to start this spring morning. I'm crouched in a caravan-sized chicken-wire pen halfway down the garden, trying to tip tinned Chappie into a plastic feeding dish. Upon my intrusion through the safety door, four orphaned fox cubs displayed the appropriate response of wild animals and bolted into their hutch in a tangle of black paws and slender white-tipped brushes. The fifth bounded trustingly straight towards me.

Chatter was the name given to her by her original rescuers; she demonstrates why with a non-stop torrent of squeals and whimpers, every inch of her orange and white frame tensed with unbearable excitement. She cannot even spy me in the garden without leaping at the wire door and climbing five feet above the ground, serenading me like a puppy anticipating a walk, before falling with a horrifying thump upon the straw. But apparently she is indestructible as well as irrepressible. And today, she has climbed not the door but me.

She scrambles up my back until she stands on my hair, chattering with bubbling abandon as I fill the dish, whistling like a soprano recorder being cleaned. She forgets me in the joy of gulping down dogfood; gradually, as I retreat to a corner, other pointed muzzles and raven-black ears appear from the hutch. The four wilder cubs take their turn at the dish, but there are no manners around this table. Rear quarters are slammed into each other, and ears flattened to the horizontal like warning flags.

Chatter.

I never tire of watching this makeshift family of five orphans. They were all found by kind-hearted members of the public in various parts of south-east England and taken to the Fox Project, a charity based in Tunbridge Wells. I don't know their histories but can surmise; cars kill vixens whatever the season, and others may be abandoned by their mothers, or simply lost. The Fox Project rehabituates hundreds of rescued cubs each year, rotating them between volunteers to prevent them bonding with any human carer. It may be tempting to cuddle a young animal as pretty as a fox cub, but if they are to survive in the wild as adults, they need to be treated as wildlife. The world is not a safe place for a fox with the trust of a dog, and I don't encourage Chatter's enthusiasm.

So I watch from afar as she leaps upon a tattered teddy bear, spinning at shocking speed towards a nylabone. She springs sideways. She springs backwards. It is as though her tiny ginger legs are made from rubber. The best toys, of course, are her adopted littermates – they zip around the pen like hopping puppies at turbo speed, brushes in the air, mouths open in the universal mammal play-face. They are foxes in the raw, expressing ancient survival behaviours hard-wired by evolution, yet needing polishing through play and experience. In appearance, too, they are not quite there; their legs and necks are too long, their coats silky-smooth, their eyes wide and their ears very big.

Play is a teacher of acceptable manners and a mock

Feeding time brings squabbles.

stage for learning how to fight and outwit rivals. This early contact helps define their sense of their own species and equips them for adult interactions. Orphaned wildlife that grow up only in human company risk being as lost in the wild as a feral child raised by Kipling's wolves would be in a London office. This is a major reason why all responsible rescue groups rear their babies in groups – and why it's so important that well-meaning people don't try to look after rescues in a corner of their kitchen.

Food is a part of their games; they pin each other with leaps that will one day secure a hapless vole. Such a curious thing to see all the familiar fox behaviours in miniature, from swishing brushes to caching morsels – at present, peanut butter on toast – under the straw. But no living creature stays young forever. And one day, they are driven to a pen in the grounds of a large property in Kent, a bit taller now, with more of the wild in their amber eyes. After some nights there, the door is left ajar. The real world is theirs to explore.

THERE ARE TWO great myths about the fox's social life. The first is that it doesn't have any – it is not hard to find old natural history guides referring to them as solitary, the product of sincere but incomplete observations. The second is that they are small ginger wolves, hunting prey in fearsome packs on dark city streets. The reality is more interesting. We have already considered the fox's interactions with the physical landscape, but the most important lines on its map are written in the scent of other foxes. They are a paradox: independent and interdependent, each fox one self-interested thread in the vast tapestry of the vulpine population. They are creatures of sociality and strife.

A fox family – officially known as a *skulk*, although the term is seldom used – is fundamentally different from a wolf pack. Wolves are intensely cooperative creatures that develop such strong emotional bonds that some researchers claim to have identified a specific 'mourning howl', uttered when a family member is dead. When you hunt bison or moose – animals that require days of travel to locate, and can easily kill you – having a close, cooperative family brings some very real survival benefits.

What if you hunt voles?

The fear of some urban commentators that foxes are about to 'pack up' are defied by the physiology of the species. There is no evolutionary pressure on foxes to develop cooperative hunting, because if three foxes pounced on the same rodent, two would go hungry. There have been rare observations from Sweden of two foxes attacking the same roe deer in deep snow, but without evidence that the onslaught was coordinated. It's unlikely that it was. The brains of wolves and other group-hunting canids are structurally more complex than those of foxes. They contain a dimple in the coronal gyrus (in the frontal region), and a significantly larger prorean gyrus. Although difficult to test, it is believed that these additional brain folds relate to the wolf's sophisticated social behaviour.

But foxes do not walk alone. On the contrary, they are acutely aware of their own kind. Their interactions are often fraught, because encountering another fox is a mixed blessing; it may be a mate, or an ally in defending a common territory, but it is also a potential competitor for breeding rights and food resources.

Roughly, foxes can be divided into four social classes: breeders, subordinates, juveniles and transients. On a map of your neighbourhood, draw a circle and label it 'Fox Family 1'. In this group, perhaps, there is a mated pair and cubs of this year. Draw another circle which slightly overlaps the first. This is 'Fox Family 2'. Again, there is a mated pair and their cubs, but there is also a yearling vixen who has not reproduced. She is a subordinate, helping to raise her younger brothers and sisters. Finally, write *transient* wherever you like. There are many foxes, especially young males, who have neither territory nor family.

This is the simplified political world of the fox. Next time you see a fox trot across the road in front of your car, consider it as a character in a play, different in social status from the other foxes with which it interacts, yet irrevocably connected to them.

WE LOVE THE fox's guile, agility and grace. It is hard to believe that at the start of its life, this cryptic, super-alert athlete, with some physical senses superior to our own, is a blind and deaf ball of chocolate fur.

Imagine a leafy west London suburb in mid-March, when daffodils brighten the roundabouts and roads are a mess of traffic cones as councils desperately try to spend their surplus budgets on roadworks before the end of the financial year. In a small beech copse behind some grand Victorian houses and their gardens with trampolines and windchimes, a vixen rests after a 53-day pregnancy. She has given birth in one of the many dens – called *earths* – that she dug while her litter was growing inside her.

Many foxes take their first breath in accommodation inadvertently provided by people. We build sheds for lawnmowers; vixens find a warm den underneath, cheering Facebook's garden wildlife pages. Our ancestors introduced rabbits into these islands, and today's foxes sometimes enlarge warrens into vulpine earths. Our railway banks provide a fine slope gradient for digging. Our graveyards are quiet, and often have overgrown corners.

Apart from our help, there is also the badger, the greatest digger of all our wildlife – many creatures exploit their roomy setts. Rabbits and mice are often tenants, and foxes too will move inside. Interestingly, in the wildwoods of Poland, foxes tend to select dens near to the woodland edge, and often close to rivers, pastures and meadows – all good hunting sites.

Our vixen in west London is a relative veteran at the cub-raising business; she has just reached the grand age of four. In some populations at least, younger vixens have larger litters with up to ten being reported on rare occasions, but even so, early mortality is very high. In London, an average of 3.9 cubs per breeding female survive to the six weeks milestone. Many will die before the year is over.

The vixen cleans the newcomers. Except for the white tail tip, they are entirely covered in short dark brown fur, and so tiny that all four together weigh about the same as a pint of milk. They cannot regulate their own body temperature, so the vixen does not travel far for the first few weeks. Her mate brings her a pigeon he has caught in one of the gardens, and the family's 'helper' – a subordinate vixen – also gathers scraps. Three adults is quite a small family; it can be six or more where foxes are deliberately fed by householders.

Youngster aged three to four weeks.

The cubs are too young for any sustenance except milk. All wildlife rehab workers know that cow's milk gives most young mammals diarrhoea, and goat's milk is more digestible, but ideally foxes should drink fox milk. It contains about the same fat levels as human breast milk, but its protein content is far higher: 6.7 per cent compared to 0.8 per cent. The cubs grow with its nourishment; their eyes open with grey-blue irises, their ears begin to unfold, and their fur gradually moults into the orange-and-white pelt so instantly recognised by humankind. After four weeks of life, as bluebells turn the woodland floor into a rich cobalt carpet and the church down the road celebrates Easter, they venture into the light.

Now the challenge of keeping them alive takes on a new dimension and, at least in some families, duties are shared out. In Ontario, male foxes guard the den area, while the vixen visits the den itself for longer. But all three adults forage this family's home range, which in an urban environment can be as tiny as 8.5 hectares – about six football pitches.

The subordinate vixen finds food deliberately provided for her by an old lady in one of the Victorian houses and jams her jaws with bones and biscuits, running at speed back to the den. Why doesn't she simply gulp it down and then regurgitate it as 'baby food'? Wolves do. But despite a persistent rumour that foxes mimic their large cousin's pup-feeding behaviour, proof remains elusive, even though countless thousands of garden wildlife watchers avidly observe vixens raising their cubs on the lawn each year. A wolf's insides include a large stomach to accommodate its dramatic feast-or-famine lifestyle. Foxes feed little and often. Their stomachs are small.

So cramming everything into their slender muzzles is a better plan. They wedge in so much that they resemble a puffin carrying sprats to its chicks. Each fox requires about 2.5 kg of food a week to survive, and a breeding vixen needs an extra 0.6 kg – about 15 voles – per cub.

The youngsters squabble over the scraps that the subordinate has provided, unaware that she was nearly a mother herself. Either she lost her own cubs in late pregnancy – as is common in subordinates due to the reduced prolactin levels caused by the stress of their low social status – or the dominant vixen killed them at birth. It sounds brutal, but nature is selfish, and no vixen wants her cubs to have competitors. Infanticide is common in the wild. Lions and tigers are particularly notorious.

Even for the dominant vixen's cubs, this is a dangerous time. Mange from parasites, cars, domestic dogs and cats – risks are everywhere. She may move them if disturbed, or bluff-charge predators, even dangerous ones. Jack Russells are rough-tough little terriers that were originally bred to battle foxes in their earths, but that didn't stop one vixen charging a terrier owned by a friend. Of course, had it come to an actual fight, the dog would easily have won – but bluff-charging is amazingly effective. Elephants may not really be afraid of mice, but something small dashing frenziedly towards you is likely to give anyone pause for thought.

By early summer, the vixen looks bedraggled. The

exhaustion of motherhood results in a tattered coat and battered brush, and householders dismiss her as scruffy. As a human who has fostered ten fox cubs, I have a lot of respect for wild vixens. In my experience, the average cub has the energy of ten dog puppies. These youngsters tear around in circles like orange whirlwinds. They leap, climb, scrap, bury, undo shoelaces and get themselves into all kinds of mischief. It's no wonder that vixens benefit from the 'auntie' support of subordinates.

Or do they? Strangely enough, while helpers guard, groom and feed the cubs, they appear to make no real

In cooler climates cubs, like these in Canada, have thicker coats, but an equal urge to play.

difference to the youngsters' odds of survival. And their own longevity is sharply less than that of a dominant fox; in one study, dominants lived an average of 4.5 years, while subordinates survived for only two. So why all the effort?

Subordinates may glean three advantages. Firstly, they are at least in a familiar territory, rather than dispersing through potentially dangerous terrain. Secondly, they may gain experience through their 'nannying' that will benefit them if they should eventually have a litter of their own. And thirdly – and probably most importantly – they are in prime position to take over the territory and breeding rights should the dominant vixen die. None of this is

Subordinate adults readily assist with cub-raising duties.

really convincing, however; subordinates only achieve the same likelihood of breeding as foxes who establish a new territory.

Regardless, the cubs keep growing throughout the summer, and by the time August brings its hosepipe bans and rainy bank holidays, they have acquired an unmistakable teenage gawkiness. Their legs are long – far too lanky for their bodies – and their ears seem too big. Their coats retain a slight softness, and, in males, skulls and shoulders are narrower than those of mature adults. They begin to wander alone or in pairs, hunting simple prey like earthworms. But as the shorter nights arrive, so do vulpine hormones. Many wander away from their parents for pastures and partners of their own.

Every young fox meets this junction: to risk dispersing into the unknown, or remain for the questionable reward of subordinate life. Interestingly, foxes in one US study dispersed earlier and further if they were heavier – which could be translated into having better fat reserves or a competitive advantage in fights with other foxes. Country-born cubs also left their parents before their city counterparts, although they did not travel as far. Dispersers had a tendency to head northwards – perhaps another tantalising hint of their apparent ability to perceive the geomagnetic field.

But most do not travel great distances, especially in areas with high fox densities, although males venture further than their sisters, and the odd individual undertakes a prodigious odyssey; one American fox dispersed over 400 km (248 miles). Not to be outdone, British foxes have a champion in Fleet, a famous male radio-collared in Brighton, who managed a meandering hike of 315 km (195 miles). The males are seeking vixens; the plaintive barks of courting foxes echo through the bare November woodlands, and many are hit by drivers. Road mortality is high during the breeding season.

The survivors try to trail a vixen, but while they are ready to mate, she is not. Remarkably, she is only fertile for three days a year, usually in the week after Christmas. They may mate repeatedly at this time, and pairs stay closer together while hunting and resting than other fox groupings. Even so, it doesn't mean that a litter will have only one father.

Foxes are often described as monogamous, but genetic studies have shown a more complex picture. Clearly there is some motive for males to risk fights by intruding. The legacy of such encounters are the tooth-mark scars on many a muzzle. The occasional male has turned up in my garden in mid-winter dripping facial blood from a particularly bruising encounter. However, only the largest male foxes seem successful at mating with females from other groups.

At any given time, many fox territories in southeast England will be home to a breeding pair, several subordinates, some cubs of the year and possibly a transient or two. Still, foxes are notorious for their behavioural plasticity – variations on the stereotypical fox family abound. And sometimes in my part of the Surrey Hills, we have been in the strange situation of having almost no cubs at all.

Six adult foxes on one sunny July morning; this is a rare crowd for one meadow. Ears poke through the grass, and where horses have grazed the vegetation into withered stumps, vulpine bodies in velvet-smooth summer coats stretch themselves, mouths gently panting. I tarry by the gate, ensuring their natural behaviour by keeping my distance – too close, and they will freeze to watch me. Minutes roll by, and still the foxes rest without sleeping, idling around one another, strolling between bemused foals.

One vixen, small, thin, yet bright-eyed, navigates past a black pony. Neither she nor any of these vixens are lactating. We now know that foxes really do practise birth control – not a collective decision to deliberately restrict population growth, but a regime enforced by the dominant vixen. Is there an additional switch, where the density reaches such extreme proportions that no cubs are born at all? Or was there some external factor – bad weather, too many fireworks – that interfered with their breeding season last winter?

The foxes don't ponder the mystery. One crouches in the grass to stalk its sibling like a cat, reminding me of another quality of fox sociality. When not fighting, breeding or feeding, they will engage in the most extraordinarily raucous play.

It can be triggered by snow. Think of a dog's thrill at tasting its first winter wonderland, and you'll picture the scene. In this very field, not long ago, four foxes hurtled about after an unexpected dusting. In my garden, an April snow shower once brought out the youth in a seven-year-old dominant male, and he play-fought with a transient vixen, wrestling open-mouthed like the cubs in the pen.

Or it can be seasonal. May seems to be particularly prone to group play, perhaps due to hormonal changes associated with cub-raising. Up to five of them have come charging around my lawn in broad daylight, rolling over, playing tag, leaping in and out of bushes. They even play king of the castle on a heap of turf.

A fox's social life is a complex and sometimes violent one. But there are moments in which they appear to enjoy not being alone.

6

The Fox and its Neighbours

WITHOUT A DOUBT, the most important animals in a fox's life are other foxes. But their world extends beyond vulpine society. Daily, their paths cross those of other species, in a subtle, interlinking matrix of living organisms that science has still not fully explored. That diagram in schoolbooks – grass eaten by rabbit eaten by fox – gives little hint of the magnificent complexity of our ecosystems. So many types of life, so many ways in which they can impact each other; it is overwhelming, a riddle with a different answer in every copse.

Those encounters can be dramatic. Foxes do hurtle across meadows to seize unwary rabbits. But they are also nudged by indirect forces. Whether a hedgerow is rhododendron or hazel, whether sheep or horses graze a meadow, whether the geology permits permanent ponds – it all affects how and where a fox will spend its days.

Foxes would not classify Earth's estimated 1.2 million wild species in the same way as scientists, even if they had the mental capacity. Their reasoning, or instinct, is revealed through their behaviour. Some species are approached because they can provide food or play. Others are observed or avoided because they are competitors or pose danger. Of course, the real world always has surprises. Notwithstanding the occasional fox who plays benignly with dogs or pet rabbits, we can at least attempt to divide the neighbours by how foxes interact with them.

Food

August brings grandmothers and their little ones to the hedgerows in my part of Surrey, gathering succulent blackberries to mix with Bramley apples in a future pie. Foxes, too, scour the brambles; in this most leisurely of hunts, they wander down the chalky hillsides nibbling berries like a bear. Yes, foxes are carnivorans, but their unspecialised carnassials are a clue that their diets are more diverse than any restaurant's menu.

For fruiting bushes and trees, the fox is not a predator, but rather an ally that absorbs calories from the flesh of the fruit while safely carrying the precious seeds in its digestive system to pastures new. Glance at your local wood for a moment – how many brambles, rowan or wild cherry trees were inadvertently planted by a fox?

Mediterranean hackberry.

Seeds excreted in scat are not just a safe distance from the overbearing competition of the parent tree. For some species at least, passing through a fox's gut dramatically increases their odds of survival. Fox-dispersed Mediterranean hackberry seeds germinate an average of 25 days earlier than their uneaten peers, and are much more likely to survive.

As a final note on botany – although not dietary – foxes also influence wild plants by the act of digging earths. In the wild forests of Białowieża, plant diversity is richer around fox dens, where some species thrive on the soil disturbance and changed pH levels. In that area, the wild boar is known as the 'forest gardener' because plants benefit from its spectacular rooting, but in a more reserved and British way, foxes might also be described as contributing to woodland horticulture.

Slightly more mobile than plants are the earthworms that foxes avidly consume, weather permitting. Rainy nights are earthworm nights; on the other hand, high winds reduce the number of earthworms at the surface. A fox on an earthworm hunt meanders slowly across suitable fields, listening for the worm's movements through the soil in a foraging technique that has been compared to that of the blackbird. This is, of course, predation; the target is killed and consumed.

But while fruiting plants, so to speak, want to be eaten by foxes, and invertebrates such as earthworms aren't much of a challenge, creatures with wings are – at least, most of the time.

'BLOODBATH AT LONDON ZOO' exclaimed the *Daily Mail* in October 2010, after hearing – six months after the event – that a fox had slipped through a broken electric fence and killed several penguins. Without wasting space correcting the scientific errors in the *Mail*'s article (and alas there were many), it is true that foxes regularly kill confined birds. But the perennial tension between chicken farmers and foxes gives a false impression of their bird-hunting prowess. A bird in a pen presents no challenge. One with the whole sky as an escape route does.

In fact, birds have the least to worry about of all potential prey. A detailed Canadian study found that foxes were successful in 82 per cent of insect hunts, 23 per cent of mammal hunts, but only 2 per cent of attacks on birds. I have watched a young fox optimistically charge a wood-pigeon in the centre of a field, but inevitably the intended meal gets airborne, unless it is close enough to shrubbery for a surprise attack.

Only when birds are nesting – especially on the ground – are foxes effective at catching them. This has caused controversy, and not only with gamekeepers who have a commercial interest in spiking pheasant numbers to ecologically eye-watering densities. Foxes take some clutches laid by lapwings and other declining waders. It may seem disingenuous that the human species, whose agricultural modernisation sent lapwing numbers plummeting in the first place, should condemn a native wild predator, but reducing wader nest predation is now widely considered a conservation necessity. Lapwing chicks appear to be a chance catch rather than something that foxes actively seek, and research in Norfolk has given some hope that boosting vole populations through habitat manipulation can divert predation away from waders.

Other food species are exploited as they are found, including fungi, reptiles, fish and – in other countries – even wild turtle eggs. Nevertheless, it is the fox's ability to capture other mammals that resulted in that old GCSE diagram of grass > rabbit > fox.

Most of the time, grass > vole > fox would be better. Foxes are stunningly expert at pouncing on rodents. Every winter, footage circulates from North America of a richly rufous fox diving headfirst into a deep snowpack to extract a lemming or meadow vole. Their long limbs, light bones, and small stomachs gift them a deft and precise pounce. Our foxes rarely encounter prolonged snow, but they deploy identical arching leaps – called *mousing jumps* – upon short-tailed field voles.

Ecologists like field voles. When you cage-trap them for study, they sit patiently in the weighing bag, round eyes watching, their scruffy coats always looking like they've had a wild night at a rock concert. They are dumpy, snub-nosed, tiny-eared little things, not prone to wriggling, and happily less dramatic than the biting, leaping ball of fury that is a captured mouse. Field voles may number 75 million in the British Isles – they are more common than people, but far harder to spot, because they live within the thick matrix at the base of rough tussocky grass. At vole-height, a normal field is a complex network of narrow runs. Here they live, fight, store cropped stems and produce tiny green droppings.

Short-tailed field vole.

Foxes like voles too, but for different reasons. In winters of high vole densities they may eat little else. Every January I see foxes aplenty in rough grassland, illuminated by the weak light of the winter sun – voles are more active in daylight during cold weather. Unfortunately, one creature that is somewhat bad news for the vole is the sheep, a native of Mesopotamia turned loose by the million into the British countryside. Livestock can eat to the bones the rich clumpy grass that voles require. Vole numbers, not unreasonably, tend to go down, and so do fox visits.

A fox on a rodent hunt approaches with stealth. Seeing a vole in snow or thick grass is impossible, but the fox tunes its ears like radio receivers, slowly rotating its head. The accuracy this provides is astounding. On many occasions, however, the fox will have to work hard for its meal. One study from Croatia found an average of seven attempted mousing leaps per hunt, and the duration of the whole predation event varied from a few seconds to well over an hour.

As an aside, foxes are one of the few British species capable of catching adult rats. Many people cite this as a positive of having wild foxes as neighbours.

Bigger still are rabbits and hares, yet only one of our three species – the enigmatic mountain hare – is a true British native. The abundantly familiar rabbit roams our countryside courtesy of our Norman or possibly Roman ancestors. They were originally introduced as human food; today, many wild predators feed on them instead.

'I have a rabbit's view of ferrets,' Dr Diana Bell, my university supervisor, told me as our class examined various mammalian relics: whale vertebrae, a stuffed fox, and here, a ferret's skull with pointed canines. Seeing the world from a rabbit's perspective means acute awareness of predators, including foxes. Diana's famous rabbit research at the University of East Anglia straddled the fox divide. In her early years of counting and measuring bunnies in the loosely fenced rabbit research meadow, Norfolk's foxes were few and far between. Rabbits fell victim to kestrels, stoats and myxomatosis, and students sat in the rabbit-watching tower to take notes.

But the foxes came back. By the time the BBC *Springwatch* crew took up station in the tower, foxes were resident on campus once more, and the film-makers caught some fine daylight footage of a large dogfox eyeing up the research subjects. Do foxes actually limit rabbit population growth? That is still uncertain; it is hard to tease apart the roles of predation and myxomatosis.

Once we get above hares on the size scale, a fox's prey options might be thought to reduce. But despite popular belief, British hooved mammals do still have a natural predator. To be clear, a fox that tries to run down a deer like a lioness will end up hungry, exhausted and possibly injured. An adult deer is a self-defence maestro; a well-placed kick can instantly kill a wolf, let alone a 4.5 kg (10 lb) fox. Nor are deer the gentle pacifists of Disney fable. In the Canadian Rockies, they attack far more people than do bears, and don't take kindly to threats in the UK either. I once saw a roe buck startled by a vole-hunting fox, and the deer bounded off a few yards, circled back and charged the surprised fox right out of the meadow.

The deer that fall to foxes are nearly always very young, but in some places, foxes take plenty of them. A long-term study in Sweden found that up to 81 per cent of roe deer fawns born each year were killed by foxes, and they were mostly predated before they were a month old. Young sika deer in Ireland are also targeted, with up to 50 per cent taken. A few adults have been recorded as fox prey in Scandinavia during severe snow. There is substantial evidence from Sweden that foxes are a limiting factor on roe deer numbers – that is, they slow down the growth of the entire population. The deer-troubled forester, or indeed the many conservationists worried about overgrazing in woodlands, should welcome the presence of foxes.

Of course, the safest way to eat deer is when the animal is already dead. About a third of fox diets in Białowieża forest is wild boar and deer carrion, much of which is leftovers from wolf and lynx predation, and weather-killed red deer are exploited by foxes in the harsh Scottish Highlands. Deer are also frequently involved in road accidents, especially in south-east England. When two deer were hit close to my house in early 2015, I decided to use the opportunity to research fox scavenging behaviour.

The results were intriguing. Camera-trap footage shows the first fox on the scene biting the deer repeatedly, but without breaking through the tough hide. It paws it, yanks it, but to no avail. Astonishingly, it then seizes the carcass – which is at least twice its own weight – and hauls it ten yards over the woodland floor, its body tensed like a dog tugging a rope. One of the few published studies on English fox use of deer carrion, from Dorset, found that they repeatedly attempted to remove carcasses weighing up to 24 kg (52 lb) from the site of discovery. Hypothetically, it may be a 'ghost behaviour', an instinct hardwired from the days when they competed with wolves, bears and wild boar for carrion – although I saw no attempt at concealment.

The Dorset study found that scavenging commenced after a carcass had been decomposing for over two weeks. My foxes did begin eating on the night of discovery, but only after shearing off armfuls of fur with their carnassials. Puncturing an adult deer carcass is clearly difficult for a fox. In the wilderness, large hoofed mammals are often sliced by the formidable teeth of a wolf, lynx or hyena before lesser carnivorans find them.

Rivals

Mid-October in Norfolk's glorious Yare Valley; I ventured outside with my dog into air infused with North Sea chill. Mist resting limp on the river, frosted reeds caught gold in the newborn sun – Norfolk has a dreamlike quality on early winter mornings, a surreal natural theatre in which some ancient instinct informs you that drama must soon commence. I didn't have to wait long.

On a field cropped low by Shetland ponies, a young male fox sat and stared at me. I wasn't his only observer.

One, two, three, four, five – the magpie rhyme ran short as a nine-strong black-and-white flying cloud envel-

oped him like paparazzi. He trotted across the field, and they followed. He sat by an ill-tempered pony, and they watched. He entered the stable, and they perched by the door – and by his paws, lining up within inches of his jaws to compose one of my most bizarre and popular wildlife images.

They were fixated by him. Whatever he did, his groupies gawked. At the risk of anthropomorphising, I was certain I could see pure bewilderment in his eyes. Years later, after being mobbed by spider monkeys while studying jaguars in tropical Mexico, I gained some appreciation of what it is like to be followed by small creatures with the relentlessness of the Borg and the comedic chaos of Mr Bean.

Why should a fox be irresistible to a magpie? Nothing is simple about corvids, the bird family that contains magpies, ravens and crows. They are astonishingly intelligent and playful, and undeniably a little odd. They watch and hassle eagles, humans and big cats too. My first encounter with a puma – in the dense coniferous rainforests of western Canada – was courtesy of a murder of crows, whose raging alerted me to the shy cat sheltering in the bushes.

When we talk about wildlife species interacting, the mental image is a lion leaping on a zebra. But while predation is dramatic, *competition* is arguably a stronger force of nature. Basic ecological theory states that where two species exploit the same resource, and that resource is in limited supply, one species will eventually out-compete the other and send it to local extinction. That is, unless the weaker species survives by evolving into a new niche.

Do foxes and magpies use the same food resources? Absolutely. However, there is no evidence in Britain that carrion and the like are rare enough for one species to seriously impact the other. It is kleptoparasitism; some corvid fox-watching can be explained as the magpies waiting for their victim to bury surplus food in a cache. Magpies may not steal shiny rings, but they do raid fox caches.

In harsher climates, their impact may be more substantial. Dr J David Henry's wonderful observations in the beautiful but bitterly cold forests of northern Saskatch-

Despite their animosity, foxes seldom lunge at magpies that venture close.

ewan, Canada, show just how thoroughly foxes attempt to evade the scavenging menace. He watched as holes were dug, the item buried, and then disguised with leaves. One fox even nosed snow over its own tracks as it backed away from a cache, pre-empting magpies following its trail.

Mostly, magpies seem like a slightly droll nuisance, something as inevitable in a fox's life as November rain. But the magpie has a big brother in the carrion crow, and its relationship with rivals is less whimsical. During the nesting season, crows will mob anything that appears remotely predatory, from herons to squirrels. Last spring, I saw a vixen ducking at high speed with flattened ears as a croaking black warbird with a wingspan almost the length of her body swooped low over her head. It was a bit like the Nazgûl menacing Gondor's soldiers in *The Lord of the Rings*.

A large number of British species could be described as competing with foxes. Barn owls and kestrels also hunt rodents; rabbits are eaten by buzzards, stoats and wildcats. Fruit and blackberries are consumed by small mammals, insects and migratory birds, while invertebrates also fall prey to spiders. But foxes also have a much larger rival.

IT WAS THE PEANUTS that did it. A luxury mobile home in a south coast caravan park might not be Earth's greatest wildlife hide, but its pitch backed onto a sweet chestnut wood riddled with holes. Come nightfall, the peanuts that had been scattered from the birdfeeders by nuthatches brought forth the hungry. A thickly-furred vixen was first, looking over her shoulder as she crunched the nuts. She was nervous – with good reason. One badger weighs as much as three foxes. Six badgers means the game is over. When they came, her eyes grew dark and her ears sank as flat as paddles – and she left them to their singularly noisy feast.

Exactly how badgers depress fox numbers is not yet understood, but where they have disappeared from the countryside – as in the government's ill-advised TB culls – the fox population has locally more than doubled. Could this be competition? After all, foxes and badgers have a considerable overlap in their diets and, unlike magpies, a badger is strong enough to drive foxes away from a feeding opportunity. Yet they are often filmed together in gardens, and foxes sometimes even raise their young in active badger setts; studies have shown a considerable amount of indifference between the two species. Perhaps badger predation on earthworms makes a critical difference to the survival rates of fox cubs, which often exploit this resource. It's an interesting area which deserves more research. Regardless, no other European species has ever been demonstrated to have such an enormous impact on fox numbers.

Danger

It may be a surprise that fox populations are predicted to rise as wolves gradually recover their historical range in the United States. It is widely but incorrectly assumed in Britain that reintroduced wolves would cause foxes to dramatically decline, a feat which dedicated pest controllers often fail to achieve.

Wolves are not fox-eating machines. They are generally big game hunters, and on the European Plain they subsist on the continent's largest mammals: red deer, moose and wild boar. As a general principle of calories and physics, wild carnivores that weigh over 21 kg (46 lb) tend to hunt creatures bigger than themselves, unless, like bears, they're shoring up their diets with energy-rich vegetarian options. With apologies to Farley Mowat of *Never Cry Wolf* fame, wolves cannot really survive on mice. Catching dozens of small animals is economically far less efficient than a single hunt that brings down a moose. Only humans think that wolves are preoccupied with foxes; wolves themselves appear to be much more interested in deer.

Not surprisingly, then, foxes aren't particularly afraid of wolves. Adolph Murie, the founding father of wolf research, investigated their relationship as long ago as the 1940s, and concluded that they could coexist in good numbers. He did discover a certain degree of mutual theft – foxes need no second invitation to help themselves to the leftovers of wolf-killed deer, and wolves readily dig up any caches that a fox has buried. They also steal earths. Every wolf den that Murie located in Alaska's Denali National Park was an enlarged ex-fox den. Many wolf dens in Białowieża are also acquired from foxes. Next time you see a wolf on a BBC documentary, remember that it might well have been born inside a burrow that was originally the work of a fox.

While competition may be more intense in semi-urban environments or where deer and boar are heavily hunted by humans, and despite possibly shared diseases and occasional violence, wolves and their small red cousins seem to have a largely mutualistic relationship.

Europe's other native canids, such as golden jackals, do not include Britain in their range. But one large cat does – or would, if early human hunting had not eliminated it.

Twenty thousand years ago, Europe was haunted by truly gargantuan felines: cave lions, Pleistocene jaguars and scimitar-toothed cats. The lion survived into historical times in Greece and the Balkans, and the leopard still persists on the Turkish fringe of our continent, but the largest cat north of the Bosporus is now the Eurasian lynx. It is the velvet-pawed ghost of the forest, seldom spied even by seasoned guides. Most felids are reclusive, for in hiding themselves from potential prey, they are incidentally concealed from their army of admirers and photographers. Some years ago, I was studying plains bison in southern Canada with camera traps, and by pure chance caught a photo of a fine bobcat – the lynx's small cousin – standing by a river on a hazy spring dawn. It was the first wild cat photographed in that area since historical records began, and not for want of looking. If felids decide not to be seen, you won't see them.

Fortunately, highly skilled trackers who read footprints, scat and other signs, and use modern technology such as radio-collars, have provided insight into the mysterious world of wild cats, and with that, data on the

Eurasian lynx's relationship with the fox. The cat native to Britain is *Lynx lynx,* the largest by far of the four lynx species – a solitary, ear-tufted, amber feline with indistinct smudged spots and a chin ruff that falls around its head like sideburns. It appears to have survived in what is now England until at least the seventh century, and has recently made a significant recovery on continental Europe. Roe deer are its preferred prey, although it will also take young red deer and reindeer, as well as hares.

Like the wolf, lynx feed foxes. In Switzerland, their return caused foxes to reduce their winter dependence on fish, invertebrates and plant material, and benefit from roe deer carrion instead. Unlike the wolf, lynx have sometimes been shown to have an impact on fox numbers. Population surveys organised by Stockholm University indicate that Finnish foxes tend to decrease where lynx are most abundant. Over in the Baltic States, however, analysis of stomachs removed from dead lynx – an unpleasant but effective technique for determining diet – suggests that foxes are only 7 per cent of food items. It is possible that lynx do not always consume foxes that they kill. Or perhaps they take more foxes in forests where deer populations are low. Perhaps impacts are patchier where wolves – which reduce lynx abundance – are themselves at high densities. The debate over the degree to which lynx influence fox populations is likely to continue for some time.

Of course, enemies of all kinds pose a greater risk to

cubs than to adults. Vixens can be highly protective of their young and sometimes bluff-charge dogs and cats which linger too close. More frequently, they relocate their cubs to a new den if disturbed. Golden eagles occasionally prey on fox cubs, and so do pine martens.

IT'S ANOTHER FROSTED January morning. The North Downs are encased in ice, and every grass stem glitters. I'm outside early, trying to walk my dog around the abandoned golf course before crowds steal the precious peace of winter. Across the shining grasses, a dark ginger fox trots with swift steps that belie his heavy winter coat. He stops on a putting green, observing me as foxes have watched potentially hazardous wild neighbours for millions of years. But unlike a lynx or cave lion, I have an interest in his own behaviour, and can deduce his place within Surrey's natural web.

He is seeking field voles. They are active in sunlight on a sub-zero winter morning such as this.

Golfers, walkers and cyclists venture into the crystalline air before retreating to warming cups of tea. They may be shadowed by human recreation and the industrialised glare on the horizon perhaps, but foxes still hunt their voles, and the natural cycle of life continues.

7

What Does the Fox Say?

Joff-tchoff-tchoff-tchoffo-tchoffo-tchoff!

Ylvis, 2014

IT'S NOT THE MOST ratings-friendly twist for a police drama.

'Sorry, Inspector. We'll have to close this case; turns out there's no victim and no crime. The witness mistook a fox's scream for a woman dying.'

It might not score well on rottentomatoes.com, but at least the scriptwriter could defend his work as realistic. The Metropolitan Police really do receive reports from people who have mistaken noisy foxes for crime victims – one every three weeks on average, according to data that I obtained in 2015. Not that murders are the only comparison for fox shrieks. Visitors to the vocalisations video on my YouTube channel have left comments claiming variously that foxes sound like monkeys, sasquatches, baby owls, chimpanzees, pigs, seagulls, Chihuahuas, sheep, gorillas, parrots, vultures, sneezing dogs, rabid zombies, raccoons, dinosaurs and apparently one gentleman's ex-wife.

What's more, that's just their most famous cry: the frantic exploding-kettle strangled-cat yelping scream that is delivered at near-hysterical pitch. It means that two foxes are engaged in hostile contact; often quite bloodless, but spectacularly intense. It may also be painful – a fox's hearing is substantially more sensitive than ours.

It would be misleading, however, to imply that all 'fox talk' takes the sledgehammer approach. Most of their communication is wonderfully subtle: pheromones, exaggerated yawns, and changes in ear angles. Their language is surprisingly rich, for they must communicate diverse information and emotions. Nearly all foxes, at some stage in their lives, will need to inform other foxes of certain desired responses:

'This is my territory, so you must leave it immediately.'

'I am looking for a mate.'

'I accept you are in control here, so stop threatening me.'

'There is danger, so cubs must hide.'

Fans of Charades can consider how difficult it might be to express these without the spoken word. Foxes may not construct sentences, but they have a fascinating variety of communication tools. Among these are about twelve different vocalisations, mostly short and sharp; wolf-like howling is beyond a fox's repertoire. They also use semio-chemicals – chemicals that convey messages – as well as their urine and scat. Additionally, of course, they are capable of expressing emotion and intent by body language. Sometimes, they truly do say it best when they say nothing at all.

I CHRISTENED HIM the Interloper. He was a robust fox, loose-limbed like a deerhound, paler than average with striking moustache-stripes, but his pointed face

lacked the broad skull of maturity. It is hard to accurately gauge an animal's size at a distance, but he was certainly taller than the resident Old Dogfox, in addition to being younger and healthier; the Old Dogfox was then recovering from sarcoptic mange, and his brush was rat-bare to the skin. Yet the territory was his, and owners usually win contests in nature, a phenomenon known as prior residency advantage. Perhaps incumbents fight harder because they have more to lose. Perhaps learned knowledge of the territory aids them. Regardless, it was clear from the Interloper's first visit that the Old Dogfox was in charge.

The garden was a lively place that year; three vixens and the Old Dogfox were frequent visitors. They drank from the pond, slept under the conifer tree, and played king of the castle on heaps of turf. The vixens' hierarchy was subtly apparent, as subordinates groomed the dominant female and vied with each other for scraps – even during their sporadic bouts of astonishingly playful gambols. When the transient young Interloper arrived, however, frivolity became friction.

He squatted on the lawn in the full glare of the security lamp, all gangly limbs and giant ears. His origins were a mystery; as a general principle, sub-adults dispersing from their families in high-density populations tend to travel fewer miles than those born where foxes are scarce, but he could have ventured from the next street, or the next village, or the next hill. Entering a garden which had long been the domain of a large, mature male fox was not a wise navigational choice.

A face beneath the conifer's twiggy hem was underlit by the security light. The Interloper sank into the grass as if he had been physically shrunk, neck low and extended, ears flattened. He yawned; not through boredom but as a signal of agitation that is also expressed by some dogs. The Old Dogfox's massive skull and scarred muzzle loomed from the shadows. His steps were considered and catlike; the feline tone enhanced by the dramatic arching of his spine. His brush, bare as it was, pointed upwards like a flag. In the seven years that I observed him, I never saw him display any serious aggression towards a vixen, regardless of her social status. A male transient awoke a new side to his character.

Why did the Interloper's arrival cause such angst? After all, not all wild species are territorial – but the Old Dogfox risked substantial loss. If the Interloper had replaced him as top fox in the garden, his access to vixens during the next breeding season would have become more unpredictable. Outright eviction from a familiar territory into the great unknown might be very dangerous: more hazards, more rivals, and less certainty about where to find food and shelter.

From his perspective, this intruder needed to be removed. Killing the Interloper would have served his purpose, but the nuclear button approach is surprisingly uncommon in the natural world. If intraspecific conflict (competition between individuals of the same species) can be resolved without bloodletting, it often will be. Ecologists with mathematical inclinations have modelled this in the 'hawk-dove' game: a statistical exercise which shows that, over time, outright aggression is actually less profitable than a more muted strategy. Being nasty is risky. Your intended victim might be nasty back – and win, leaving you not only territory-less but also seriously hurt.

So the Interloper and the Old Dogfox settled for vulpine dialogue rather than puncturing each other with their canines. The intruder exuded submissive body language: brush curled almost between his legs, posture close to the grass. The Old Dogfox stalked about the lawn with brush high, before clamping his jaws on the lowest branches of the conifer and repeatedly yanking them with all his might. Show of strength? Spreading of scent? Displacement behaviour, akin to a human pounding the table during a heated row? It was unclear.

Their quarrel simmered on throughout the summer. Perhaps surprisingly, the Interloper chose to remain in the garden and endure daily reminders of his social inferiority. Like chronically stressed humans, foxes that experience problematic social interactions show higher levels of cortisol in their blood stream, with a disturbingly diverse impact on health.

Tension trickled outwards into the rest of the garden group. One little vixen, in particular, achieved local notoriety for pacing up and down the garden path, her scratchy barks and doleful chirps peppering the midsummer air. I asked my blog readers to name her, and she became the Chipped Vixen because of a notch in one ear. Her talkative nature provided realms of interesting camcorder clips.

It never occurred to me that her voice was fated to go viral.

FOR MILLIONS UPON millions of teenagers in 2014, only one song existed. Two handsome Norwegian lads – band name Ylvis – had taken it upon themselves to dress up in fox costumes and dance to an inexplicably catchy tune in a northern pine forest. '*What Does the Fox Say?*' they sang, pouring their vocal cords and hearts into the mystery, while fox scientists everywhere scratched their heads and considered the irony of implying muteness of arguably the noisiest wild mammal in Europe.

A roaring tidal wave of Ylvis fans shortly thereafter descended upon my YouTube video of the Chipped Vixen, and nearly all of them posted the same comment: 'Oh, so *that's* what the fox says!'

Upon my first hearing of the infamous fox song, I actually thought that the *Jacha-chacha-chacha-chow* chorus lyrics were a deliberate attempt to imitate a squabbling fox cub. Now, I'm not so sure; I did contact Ylvis's agent to clarify but am yet to receive a response. Either way, in the midst of parodying pop music, Ylvis exposed just how many members of the public do not recognise a fox's voice, even though many of them will have heard it. On the other hand, foxes are not going to be communicating with horses by Morse code any time soon. And sorry, Vegard and Bård, but your mysterious guardian angel most definitely does not keep its voice a secret, although I've met a few people who wished that it did.

I enjoyed the Ylvis craze; it made an improbable celebrity out of the Chipped Vixen. Yet, listening to my recordings of her barks now, I am struck by how many other fox 'songs' I have yet to capture.

Season dictates playlists. Midwinter is loud in fox country. Courting foxes summon a mate with a three-tone *wah-wah-wah*, repeated over and over in the dark nights of post-Christmas chill. At close range, a second courtship call, known as *gekkering* or *clickerting*, may also be heard. More guttural in tone, it has been compared to stones being knocked together.

The hysterical yodelling screams of high conflict can be heard throughout the year, but peak during dispersal and the breeding season. I have watched foxes giving this call many times, and the frantic tension can be a prelude to physical violence. The fight described in Chapter 1 – where an intruding vixen was flipped literally upside down by an exceptionally aggressive resident subordinate female – was accompanied by shrieking. So was a quarrel between other vixens under the conifer tree; both flattened themselves entirely on the grass, from chin to brush, and proceeded to scream at each other from the range of a few inches.

Most dramatically, the back of the garden has seen warfare. Two vixens danced the foxtrot there one summer night. Jaws wide and ears sideways, screams squeaking constantly, they pressed their forepaws into each other's shoulders. One vixen pushed her rival backwards as if she were rolling a barrel, forcing her down onto the flagstone path with a thump, destabilising herself in the process. The screams continued, now with the protagonists squatting, only inches apart – and suddenly freezing, staring together towards the lawn in synchronised silence. The distraction faded and the truce shattered with it; they resumed their shouting match with even greater intensity.

Astonishingly, the Old Dogfox then bounded towards them, and they instantly broke apart. Exactly what motivates a wild animal to play 'peace-maker' when there is fighting between members of the same group is debated, but dominant individuals are more likely to display this behaviour. It is easiest to explain when the go-between stands to gain from not letting either protagonist gain a victory. Fights between rutting fallow deer, for example, are occasionally ended by a third male whose social super-

Old Dogfox.

iority is protected by permitting rivals no victories, even against each other.

The Old Dogfox, however, would not have been threatened by either of the vixens being vanquished. To date, this is the only occasion in which I have observed 'peace-making' behaviour in foxes.

Other sounds help to prevent violence rather than accompany it; variations of a cub's food-soliciting cries can be heard when a subordinate adult greets its superiors, and indeed also when a captive fox interacts with its keeper. Such noises may be accompanied by submissive or cub-like body language, such as tail wagging and a low-slung posture.

If humans easily notice sounds and general manner, we often forget the importance of scent in fox communication, despite the fact that they are one of the few wild mammals readily detectable by our own weak nostrils. Some of their sweet musky odour originates from the violet or supracaudal gland, which is found about a third of the way down their brush. The name is a tribute to the 1677 description by Casper Bartholinus that it 'smells of March violets'; worthy of flowers or not, it produces a lipoprotein secretion that stains the surrounding hairs dark and is especially active during the breeding season. The composition of the anal sac also alters at this time of year. The exact role that these play in vulpine communication and courtship is still not fully understood.

Fox urine is likewise pungent, courtesy of methyl sulphides and two other volatile compounds. Areas that they have visited are particularly odorous if mixed with fresh rainfall. Exposure to increased heat also releases the odour. Remarkably, this characteristic is a weapon against criminals in the USA. Some American commercial tree plantations and universities have a problem with thieves stealing their pines and using them as free Christmas trees. The solution of choice is to spray the trees with fox urine. Outside on campus in chilly December weather, the tree is benign. If the thief is unwise enough to cut it down and decorate it in his nice warm living room – well, at least it will be a Christmas that no one's noses will ever forget.

Male foxes, like male dogs, frequently cock their legs to urinate on prominent objects at fox nose-height. I recently caught trail camera footage of an urban fox doing precisely this on the corner of a dingy, decaying wall. Shortly afterwards, as recorded by the camera, a second fox passes the spot and sniffs, its pace downgraded from a speedy trot to an observant walk.

What could it have learned from the urine of its predecessor? At the most basic level, leaving body odour in a territory makes the area smell like that individual, and enhances the impression of ownership. As the Interloper learned to his cost, the owner of a territory has a huge intrinsic advantage in contests. In fact, an experiment with placing synthetic fox urine into the wild showed that dominant males increase their activity around the scent marks of the apparent intruder. Vixens did not change their behaviour. This fits with the conventional model that territories are really the construct of male foxes; a vixen, rather like a Victorian woman, holds property only as a proxy of her mate.

One abiding myth about urine is that foxes will avoid areas where humans relieve themselves. It is often proposed that the man of the house utilises the garden to deter foxes from entering it. Unfortunately, this is nonsensical. An urban fox that breathes human scent every waking moment is hardly likely to panic at encountering more. We overestimate our importance to foxes; the only scent marks that they truly care about are those left by members of their own species.

Claims that lion dung from British zoos will miraculously deter foxes are also somewhat dubious. One of the few objective studies showing how foxes actually react to the scent of potentially dangerous species comes from Israel, where foxes share the harsh landscape with golden jackals. Seeing a live jackal did trigger avoidance of the study area. Mere jackal urine, however, had no impact on fox behaviour whatsoever.

Logically, why would it? Rodents have been shown to avoid areas that are sprayed with predator urine, but any mouse that ignores a fox courts instant death. By no standards is the relationship of foxes to bigger carnivorans so simple.

There is risk, at least from lynx, but there are also substantial benefits – in the wilderness, a significant part of a fox's diet is carrion scavenged from large carnivore kills. Persecution of lions over the centuries has shrunk their enormous prehistoric range, and presently few or no red foxes encounter the king of beasts in their daily lives. There was a time, however, when both species roamed what is now Britain, and foxes were able to scavenge the carcasses of extinct elephants courtesy of lion predation. In East Africa, even today, wildlife watch for circling vultures to guide them to fresh lion kills. If British foxes do have any instinctive memory of big cats, lion dung might act as an attractant, not a deterrent.

On the other hand, a more rational analysis of fox urine has given us Scoot, a non-toxic formula that neutralises its odour. Consistent application of this spray seems to make it impossible for the fox to successfully mark its territory, and it will generally begin to avoid the location, an interesting insight into how much importance they place upon being able to signal that an area is 'home'. As discussed above, it is a consistent pattern in nature that territory owners usually win conflicts with intruders – but the intruder has to know that it is intruding. No scent marks mean no territory is flagged. The owner's advantage is gone.

Scat and urine are often messages for other foxes, but one of their most common usages is note-keeping for the fox itself. As animals with relatively small stomachs, foxes need to eat little and often, rather than gorging themselves like dogs or wolves. Even if they do find a surplus of food, there are limits to how much they can eat. On the other hand, walking away from the leftovers means spurning food that they may require tomorrow. The solution is caching: digging a small hole, putting the item inside, and scraping mud and leaves over the top.

How does a fox with a territory of several hectares and no map recall where it has buried its caches? Foxes do appear to remember the approximate locations. Experiments with captive foxes also suggest that the burier is more likely to find its cache than another fox that happens to be passing by.

It is after the cache has been eaten that scent-marking comes into play. Despite their good memories, it appears that foxes remind themselves that the supply is empty by urinating on the site. Larger items such as gnawed bones may be marked with scats instead.

Nor is it just edibles that receive this treatment. Forsaken toys do too, a warning I once should have heeded.

I had found my shoes bitten, tossed and forsaken on the lawn – not an uncommon experience when one lives in high-density fox country, and I gathered them up and dashed for my train. After all, they were intact, and who among my urban colleagues in that gloomy south London office would imagine that I was wearing objects that had recently been in the mouth of a wild fox?

It wasn't until I was seated at my desk that a sweet, vaguely rancid, musky perfume began to rise.

My manager, tapping away on his keyboard, paused and sniffed.

'What's that smell?'

I chose silence rather than confessing the fox's deed, and grimly attended to my work.

Minutes became hours, and the pungent scent of fox urine migrated around the office. More colleagues began to comment. 'What *is* it?' my manager exclaimed in exasperation, crawling around the floor, emptying the bin.

It was a long, lonely walk to the train station.

FOX COMMUNICATION IS rarely aimed at us, and yet we have absorbed it into our cultural arts, created mythology about it, worried about it sufficiently to create deterrents, and even – as in my farcical experience – embroiled it in office dramas. But when humanity is not part of the equation, and the earthy twang of a fox's bark rises from bare woods under winter's bright stars, their voices help restore the wild world to our consciousness too.

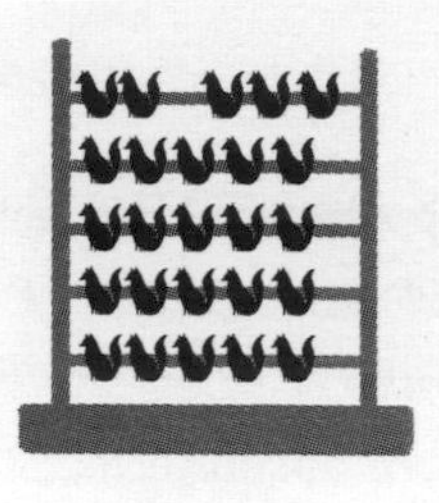

8

Counting Foxes

Perfect numbers, like perfect men, are very rare.

DESCARTES

ESTIMATE THE NUMBER of people living in your street.

I'm not sure I could, despite living in a village. Certainly I could count the houses, but without tedious hours of observation, it would be unclear if they were occupied by singletons, retired couples or middle-aged ladies with six children.

There is no faster way to upset an ecologist than to regale them with thumping claims about wildlife numbers from individuals not trained in population demographics. No sane politician would calculate future demand for housing, care homes or schools without firm, objective numbers: trends projected from censuses, and data on births, deaths and immigration. Current populations, whether they are human or wild animal, are not a matter of opinion: they are absolute facts that can be studied to a reasonable degree of accuracy using well-respected techniques.

Question 1: How do we count?

Firstly, decide the area for which data are required. From a wildlife biologist's perspective, this could be a nature reserve, a county, or even a country. Yet even in smaller sites like a town park, counting every individual is almost impossible, so we will need to take a subset of the study area. Once this has been researched thoroughly, the results can be extrapolated across the board, just as an opinion poll company interviews a representative sample to advise us about the percentage of Britons who prefer tea *sans* sugar.

But what happens if we've counted foxes in a field and 90 per cent of the park is woodland? It's unlikely that the population density is the same in both. *Stratified sampling* is clever enough to take this into account. The survey starts – like most of life's important journeys – by looking at a good map. Some skill will be required to determine the proportions of different habitats. If 10 per cent of the reserve is grassland, then you can have ten samples in that habitat and ninety in the wood.

The sample we need is a transect – one of the core units of measurement in the ecologist's world. Transects are straight (more or less) lines through a section of habitat. Their length depends upon what is being studied; for the jaguar population surveys that I led in the Mexican dry tropical forest, transects were two kilometres long, and every inch of dusty mud had to be scanned for big cat footprints. Transects walked by butterfly researchers are sometimes much shorter – but whatever the subject, they come with their own rules:

1 Transects are *regular*; all are about the same length in any particular survey.
2 Transects are *replicated*; in general, the more transects we have, the better. For a fox survey that I undertook in a local Site of Special Scientific Interest, I used 15 transects, each of which was about 500 metres in length.
3 Transects are selected *randomly*; to deliberately choose points where we know foxes are abundant will lead to biased data and an overestimate of the site's total population.
4 Transects are walked *repeatedly*, in the same conditions, to make sure our observations on

a particular day aren't a fluke, unless the survey circumstances dictate that a single sample is more appropriate.

If that sounds complicated, it's because ecological scientists are painfully aware of how important reliable data are. A hundred anecdotes are not worth an objective, unbiased population survey which has been thoroughly critiqued by the peer review process of scientific journal publishing. Unfortunately, grumbles in the gym that 'we're overrun by foxes' from a banker who spends his working life in a cosy office are not subject to the same brutal, error-killing analysis.

He might have seen the same bold fox a dozen times over; he might simply be remembering seeing foxes more this year because his mind has been turned to them by incessant media coverage. If I had been convinced that golden retrievers were in immediate danger of over-running the country, I'd probably start noticing them more too, even if absolute numbers hadn't changed.

Or perhaps he really is seeing more foxes, but only because they're now active in his street rather than last year's favoured location of the post office car park. Years ago, my garden was suddenly graced by large flocks of yellowhammers, one of Britain's most beautiful and threatened songbirds – the lemon-yellow jewels that perch on hedgerows and cry, *A little bit of bread and NO cheese*. Sadly, this wasn't a population boom; the agricultural field down the road had switched to autumn planting and deprived the birds of feeding in winter stubble. More sightings for me actually meant the population was starving to extinction. An area-wide scientific survey would have exposed the local 'increase' as a blip. And blips and anecdotes are no basis for a sound wildlife management strategy.

What is the ecologist looking for when she walks her transects? Usually, it's scat – it might be unglamorous, but at least it doesn't lie. We know how many scats foxes average per day from studies in captivity, so, with some basic maths, the number of scats in a field does give a good indication of how many foxes are present. The idea is that scats are removed when the transects are first walked, and any fresh ones on subsequent visits can be counted.

Tracks are less useful for estimating populations, because the number of footprints that an individual can leave each day is hugely variable. Actual sightings are a grey area; some foxes are shyer than others, and others appear only in the small hours of the night. There is a technique called the distance method which attempts to calculate populations from the number of animals observed and how far away they are from the observer, but the statistical analysis requires headache tablets, and in any case it's better suited to abundant and easily-spotted species like monkeys or deer.

It's also a given that no *Daily Mail* editor has ever tried to wrestle with these games of science.

Question 2: What are we counting?

A wildlife census is not simply a question of doing a head-count. Suppose you were in charge of town planning and were told that there were a hundred couples in one street. Before pouring millions of pounds into a new school, it would be prudent to determine if these people are actually of child-bearing age, or whether in fact they are all residents of a care home. Human populations, like fox populations, contain a large number of individuals who will not produce offspring in any particular year. While the overall number of foxes is interesting, the number of breeding adults is far more relevant to future population trends. Subtract juveniles and other non-breeders, and you'll be left with the *effective population.*

A good strategy for this is to count the number of litters or active dens during the spring. Again, the data will be carefully extrapolated over the wider region, taking variables like habitat into consideration.

Question 3: Why are fox numbers what they are?

London's human population is not controlled by the murder rate, and its foxes are not suppressed by human killing.

That may seem surprising. Humans are certainly capable of hunting some wild species to extinction, but rhinos, tigers and whales are large and slow breeding; foxes are neither. As for our extinct apex predators, we have already seen that wolves largely ignore them and the lynx's impact is mixed. So don't foxes have a natural predator? Will the population keep rising for ever and ever, until Britain collapses under their furry orange weight?

The fixation with the idea that fox numbers must grow unless some individuals die violently is not grounded in reality. The press seems incapable of distinguishing between *compensatory* mortality – i.e. the fox that was shot would probably have died in natural turnover before the next breeding season anyway – and *additive* mortality, where hunters kill individual elephants, for example, that would otherwise have survived. But the bigger kicker here is that predators don't always rule the world.

Perhaps it's the fault of all those spectacular television documentaries. We have been fed images of wildebeest being seized by crocodiles so many times, we have forgotten to ask whether the tail is wagging the dog. There are surprisingly few examples of predators holding prey species well below the carrying capacity of the habitat. Sometimes, it's the other way around: predator numbers are a product of prey numbers. If you have an Indian forest with lots of chital deer, you can expect your tiger population to rise.

There can also be a see-saw effect, seen most dramatically in Canadian lynx, which have population boom-and-bust cycles just like the snowshoe hares that they consume. If humans kill off the prey altogether, as the commercial bushmeat trade has tragically achieved in much of southeast Asia and central Africa, predator populations collapse.

Politicians who suggest that shooting urban foxes controls the population display their scientific naiveté. Deliberate human killing is only one facet of urban fox mortality; disease, cars and other foxes play their own roles. Nobody would claim that New York's mafia kill enough people to reduce the human population of that city, or that flats thus vacated aren't soon occupied by new families, and yet that is analogous to Boris Johnson's logic in arguing for an urban fox cull.

Foxes are more than raw numbers. Their societies are complex, with breeders, non-breeders, and juveniles. If a pest controller shoots a non-breeding adult, it will obviously make no difference to the number of cubs born the following spring. If he kills a breeding adult, a non-breeder will immediately move into the vacant niche. If he kills an entire family – as happened recently in my area with illegal poisoning – formerly transient foxes will take over the territory, often within days. The wider fox population can sustain huge annual death rates without experiencing any long-term decline.

Even if the ethical side is ignored, it's simply not practical. Many foxes are on private land whose owners wish to protect them. Nor are they easy to kill; shooting is dangerous in urban areas, and catching them in cage traps is time-consuming and ineffective. Gassing, poisoning and leg-hold traps cause terrible suffering and are strictly illegal. London borough councils stopped financing fox destruction not because they are closet animal rights activists, but because they understood it made no financial, logical or ecological sense.

So what is controlling the fox population?

Consider the inverse question: how does it grow?

Each spring, the fox population rises by hundreds of thousands when tiny blind cubs are born. Anything that impacts the flow of newborns has the potential to regulate the fox population. And nothing does that more effectively than other foxes.

No vixen cares about boosting the population of her species. It is her own genetic material – her cubs – in which she will invest every drop of energy. Other foxes and their cubs are rivals; to her, for dominant breeder status, and to her offspring, as competitors for food and subordinate attention. If another vixen in her territory produces young, she will usually try to kill them. In fact, it frequently ends before that stage, because the stress hormones in a lower-ranking vixen make it unlikely that she will carry her cubs to term.

The fierce selfishness of this social system – technically intraspecific competition – locks fox populations at a density far lower than any external force ever consistently could. England's foxes are not a uniform army where endeavours are coordinated by a major-general. Their population is better described as a hundred thousand warring tribes.

Of course, over time there will be trends. Disease can temporarily reduce local populations; sarcoptic mange caused a sharp decrease in Bristol in the 1990s, although sporadic outbreaks in my Surrey village have had little impact. Abundant food can dramatically shrink territory size while increasing the number of subordinate adults per group, effectively raising fox densities. Afforestation, removal of hedgerows and overgrazing by sheep all make habitats less suitable for foxes; conversion of some forest types into meadow may improve it. Abundant lynx or badgers can reduce fox numbers; high wolf densities can increase them. Nature is never static.

Question 4: How many foxes exist?

There is no global estimate of the red fox population, which extends from Alaska to Australia, via Saudi Arabia, Belarus and Tibet. Professor Stephen Harris and his colleagues at the Mammal Society have estimated the British population to be 240,000 in 1995. Alternatively, the Department of the Environment, Food and Rural Affairs landed on 430,000 through modelling habitat availability.

Question 5: Should I trust newspaper claims about urban fox numbers?

No.

9

In Sickness and in Health

IN STUDYING WILDLIFE, you will see wild death. Most ecologists spend a considerable proportion of their careers examining carcasses: roadkill, diseased, shot. In the wildwood, flesh is recycled, detritivores nourished, and the ecological cycle sustained. A deer that falls to a lynx gives life to martens, crows and foxes. A spruce tree that is hollowed out by pine beetles attracts three-toed woodpeckers. But when humans become actors in this cycle, the unnecessary suffering jolts.

I FOUND THE first fox stretched out on a pavement adjacent to the main road – superficially, there was nothing remarkable about that. Road accidents are no more natural than shotguns, but reducing them requires landscape planning and enhanced driver awareness. There is rarely any deliberate malice. Cars kill wildlife by the million, presenting one of the most expensive and intractable conservation and welfare issues that we are faced with today.

Only, there were no cars so very early on a Sunday morning, with fog hanging like a blanket spiked by tree trunks and lampposts. And the fox was not on the road but pushed against a wall. Nor was it the limp mess typical of roadkill. Its spine was arched, and its limbs were held high, level with its head and brush, as if it had been frozen in mid-bound.

A few hours later and twenty yards away, I stood by a second fox, also abnormally contorted. It lay next to a black bin-liner in which it had apparently been dumped.

I rang Natural England's poisons hotline, only to be stifled by a disinterested officer complaining that his budget was too low to let him investigate. But the toll kept mounting. A third fox was thrown into woodland further down the same footpath. The fourth followed soon after.

Despairing, I took the last carcass to a local vet and paid for a necropsy. A little girl tended lovingly to her dog as I carried the bagged remains of a small wild canid into the consulting room – she looked up with a horror of a question in her face, and I did not want to answer.

Its chest was full of blood from severe internal bleeding, the classic symptom of rodenticide. Deliberately or accidentally, someone in my village had illegally killed at least four foxes and thrown their carcasses out like garbage.

Wildlife crime is a crime. Those who thus indulge risk exciting interest from the criminal justice system, and a maximum penalty of six months of imprisonment – or even more if their case is referred to the Crown Court. Most police forces have at least one wildlife crime officer; the Met has an entire team. A 49-year-old man who recently crippled several foxes with gin traps in a residential area of Guildford discovered this to his cost. Foxes are protected from unnecessary cruelty under the Animal Welfare Act 2006. In addition, the law is not ambiguous towards leg-hold traps, strangling snares and poisons: using them on foxes is banned.

Even for rats, it is not permissible to simply leave poisons in the environment. Baiting that puts non-target species at risk is illegal. Unfortunately, what the law does not cover is the retrieval of poisoned rodents. If you feed anticoagulants to mice, you put everything that feeds on mice at risk. Secondary exposure to poisons is not necessarily fatal but has been linked to compromised immune systems and higher risk of disease in some species. Many foxes now carry trace amounts of rodenticides in their livers.

But while the slow smothering of the environment by the toxins that underpin our modern lifestyles is lamentable, it is deliberate poisoning that causes particular fury.

Granted, there are cases where people assume that a dead fox has been poisoned when objective evidence indicates that it has simply been hit by a car, but its frequent appearance in casual conversation has disturbed me. Incredibly, many people still seem to think that foxes are legal targets.

Should we look to the press for an explanation? Pro-hunting opinion columns and letters often remark that poisoning will increase due to the ban on hunting with hounds. Seldom do these articles point out that poisoning is just as forbidden as hunting. Whatever the intentions of the editor, these comments tacitly encourage criminal acts.

Free speech may be the mark of a democracy, but even in these days of social media, speech is absolutely not financially free. The tabloid press has a vast, expensive, well-oiled machinery through which it disseminates misinformation about the legal status of wildlife. It is not realistic to expect scientists and wildlife rescue groups – both of whom are chronically short of funds – to compete.

Just how widespread is illegal killing of foxes in Britain?

Hard statistics are difficult to obtain. I raised Freedom of Information requests with both Surrey and Metropolitan Police, but the results were incomplete in Surrey's case, and restricted to actual investigations for the Met. Only four of those were on record since 2010, which is startling. Surrey offered none, despite the fact that I had reported one myself. Despite these very limited data, it does not appear that there was a major increase in poisoning incidents following the well-publicised urban fox bite events – or if there was, few witnesses logged it with the authorities.

But it would be misleading to imply that most foxes ending up in the blankets of a wildlife rescue centre are deliberately mistreated. Many are accidental victims of the dangerous apparatus that modern humanity has introduced to the landscape: cars, broken glass, barbed wire, or even the netting of football goals. Others suffer from natural ailments, but a human has made a decision to intervene.

Many fox health issues are of the parasitical kind, which from an ecological perspective is just another type of natural interaction, like predation, competition and mutualism. Most species of parasite do not kill their host, but rather exploit it continuously. It could be said that this is far more efficient, if less admired by documentary-makers, than actual predation; a lion that kills a gazelle has no choice but to repeat the action next week, but the lion's fleas may complete most of their lifecycle without needing another victim.

An inordinate number of published scientific papers about foxes revolve around parasites. The tapeworm *Echinococcus multilocularis* is a particular favourite because of its health implications for humans, although it is not present in the United Kingdom. It is transmitted to predators via infected rodents, and foxes are by no means its only host; wolves, coyotes, and domestic dogs and cats can also carry it. If a person does catch it, for example through poor hygiene after handling contaminated dog faeces, surgery may be required to remove tapeworm cysts from their liver. Without treatment, it can be fatal. Human infections are still relatively rare, although increasing.

Various efforts have been made to reduce *E. multilocularis* in continental Europe by baiting foxes with worm-killing medication. It also appears that foxes that consume human-sourced food, such as rubbish and deliberate handouts, are less likely to be infected than those reliant on their natural diet of rodents.

Should Britain be worried about *E. multilocularis*? To be frank, no; extensive and ongoing monitoring has failed to find a single infected fox in the UK. As a side note, the related *E. granulosus* tapeworm is endemic in sheep. Ironically, the only recorded case of a human becoming infected by it in Britain was a fox hunt employee in the West Country, who ingested tapeworm eggs while feeding sheep offal to foxhounds.

But another disease is feared far more than *E. multilocularis*. I've thought more about rabies than most people; nine years ago in Canada, I was hauled into a charming rural hospital for an emergency jab with rabies serum. The cause of the alarm was a little brown bat that had become trapped in my rustic mountain cabin. I've no reason to suspect that it was rabid, or, indeed, that I was bitten by it – but it is theoretically possible to catch rabies from inhaling bat saliva in the air, and the Canadian authorities sanctioned no chances. I was booked a date with the largest needles that I've ever seen, or ever wish to see. To be fair, the rabies jab is not as painful as legend would infer, but the full course plus the serum equates to seven injections in five weeks, and even that doesn't give immunity for life.

Classical rabies – the species which is responsible for most cases in humans – is an RNA virus of the Lyssavirus family, which also includes rarer forms restricted to bats. All mammals are potential victims; infection of birds has only been recorded in laboratory conditions. It is transmitted through the blood or saliva of an infected individual, and causes severe inflammation of the brain. Medically, there is no middle ground. If a bite victim is given serum while the virus is still undergoing its long incubation, they will survive. If they wait until symptoms materialise, death is inevitable.

Rabies is a zoonotic disease; that is, one spread between animals and people. Foxes are one of several wild carnivores considered important wild rabies reservoirs, but actual transmission is not their doing – the World Health Organisation estimates that 99 per cent of human cases are due to bites by dogs. Even during rabies' pre-vaccination heyday, rabid foxes were very rarely reported to attack people, unlike infected dogs and wolves. Today, among people, fatalities are extremely rare in North America and Europe, and generally involve people who have been infected abroad.

Despite the fear that the name still evokes, rabies is one of Europe's greatest medical success stories – a lethal, endemic, contagious disease wiped clean off most of our continent's map. Apart from a different form restricted to one species of bat, rabies was declared extinct in Britain in 1922, and the last indigenous human case was in 1902. I meet people who fear British foxes for many reasons, but the occasional mention of rabies is the most inexplicable. We simply do not have this virus, and neither do our foxes.

On the continent, a mass campaign of vaccination has been spectacularly triumphant. Almost two million square kilometres (722,004 square miles) in 24 countries have been targeted with oral rabies vaccine in bait, inoculating enough foxes to break the contagion cycle. It is one of the most successful campaigns against wildlife disease ever undertaken.

France, Germany, Belgium, and mainland Spain are among those countries now classed as rabies-free; Italy, Estonia and Hungary are very close to eliminating it. It remains endemic in Turkey, Russia and Belarus, affecting their neighbours. There were 257 reported cases in Polish animals in 2012, for example, of which 77 per cent were foxes. Even so, the practical implications for anyone living or travelling in rural Poland are minimal.

IF RABIES IS THE monster that is largely in chains, sarcoptic mange is a real and constant agitator. 'Mangy foxes' has become a common insult, although many apparent victims are actually quite healthy and are simply undergoing the spring moult. Even more ironically, it now appears that *Sarcoptes scabiei,* the mite responsible for most canid mange, was originally gifted to the wild world by humanity. While this theory is disputed by some authors, others claim that mange spread from us to our domestic livestock in antiquity, and from them into wildlife. Today it is present in at least 104 species, including critically endangered African wild dogs and Ethiopian wolves, as well as pigs, llamas, and lynx. It still occasionally infests the human species, its original host, where it is given the marginally less socially weighted medical title of scabies.

It is a mite of many variants, however; as it has taken advantage of more and more species, it has evolved specialised strains. As a consequence, the form which affects foxes and dogs (*var. canis*) is of limited interest to human medicine. On the rare occasions it encounters humans – usually via domestic dogs – it usually causes nothing worse than a short-lived itch, and quickly vanishes. The form which causes scabies itself is *var. hominis,* and is only spread through contact with an infected person or their clothes.

S. scabiei is an assassin in miniature. The female is under half a millimetre (0.02 in.) in length, and the male is half as small again. It is an eight-legged invertebrate like all members of the class Arachnida, which also contains spiders, scorpions and ticks. Its life cycle runs from egg to larva, and then from nymph to adult. All four stages can take place on its unfortunate host, who may be first encountered through contact with an already infected fox, or simply in the environment – lying down in a den where another fox dropped mites, for example.

Once on board, the mites will burrow into the fox's skin, creating tiny tunnels littered with faeces, digestive enzymes, and eventually eggs. The fox's physiological response depends upon its immune status, previous exposure to mange, and the virulence of the mites that have parasitised it. Many exhibit what is known as hypersensitivity – an auto-immune reaction that may eventually kill most of the mites, but leaves the skin crusted and liable to fatal infections. Other animals build up huge numbers of mites, up to 5,000 per square centimetre (0.5 in. square). Fur loss can be severe, especially on the rump, brush and head.

Mange can kill, and sometimes in large numbers.

Fox with moderate mange on its brush and flanks.

It is not the mite itself but secondary infections – plus hypothermia, starvation or dehydration – that are fatal. Additionally, severely infected animals often fail to breed. Nevertheless, not all outbreaks result in the 95 per cent population drop temporarily witnessed in Bristol in the 1990s, or the catastrophic declines in Sweden and Illinois in earlier decades. Mange has been continuously present in south London and surrounding districts since the 1940s, without reaching anything close to epizootic levels. A perfect storm for a mange epidemic includes a high-density population without prior exposure, many individuals with compromised immune systems due to poor habitat or social stress, and a new mutation of the mite.

A fox with mange is a distressing sight: hairless, limping, disorientated and with thick grey crusts cracking its skin. It will probably be dehydrated from the raised temperature caused by secondary skin infections, and may visit ponds for an urgent drink in the middle of the day. It is not surprising that mange is the subject of many calls to wildlife rescue centres. It can be treated with the parasite-killers Ivermectin or Selamectin.

Several of my local foxes have been successfully treated with the medication mixed with a little mouthful of warm cheese, but there is a caveat. Actually, two. Firstly, while territory owners move on fairly predictable circuits and can be thrown regular doses, transients and dispersers are far more challenging. Secondly, even if the sick fox is sitting on the lawn, you still have to make sure that the medicine ends up inside it, and not into its relative who has been quietly watching from the bushes.

Sometimes, the only option is to capture the infected fox and feed it medication in the safe confines of a wildlife hospital. Unfortunately, catching foxes is easy to say and very hard to do. Leg-hold traps are illegal; tranquilliser guns look fun in computer games but are rarely useful in real life because the drugs take too long to work – a darted fox may be several streets away before it succumbs to the sedative. The most realistic technique is to lure the fox into a cage trap, a large box whose door is tripped when something seizes the bait.

I knew one of those foxes who needed to be trapped. She was a small subordinate in the Garden Group back in

2006; she projected a confident demeanour, but her lowly place in the hierarchy was never in doubt. Whether stress had impaired her immune system, and, if so, this was a factor in *S. scabiei* taking an aggressive hold of her while the higher ranking females remained in good health, is unknown. Her fur vanished from her flanks and head, her skin crusting into painful lesions. Feeding her Ivermectin in the presence of so many dominant foxes was impossible. So London Wildcare lent us a trap.

Sure enough, we caught a fox. But it was the Old Dogfox, the undisputed leader of the Garden Group. Naturally, we released him, but before long he was captured again. Ecologists who live-trap animals for scientific studies refer to some individuals as 'trap-happy'; they have learned that entering a trap means reliable food and no serious harm. Unfortunately, there is no technique for priming a cage trap to only box a particular fox. As the days drifted by, the little vixen's health deteriorated, and still she did not enter the trap. Meanwhile, a transient spotted the contraption in the garden and climbed on top of it, standing there in the glow of the patio light like a vulpine monument.

Finally, we caught her. In the trap, crouched dark-eyed against the approaching humans, she was a reminder of how small and vulnerable foxes really are. Volunteers from London Wildcare took her to their centre, where her excruciatingly itchy and sore skin was bathed in aloe vera, and the life-saving Ivermectin fed to her. A month later,

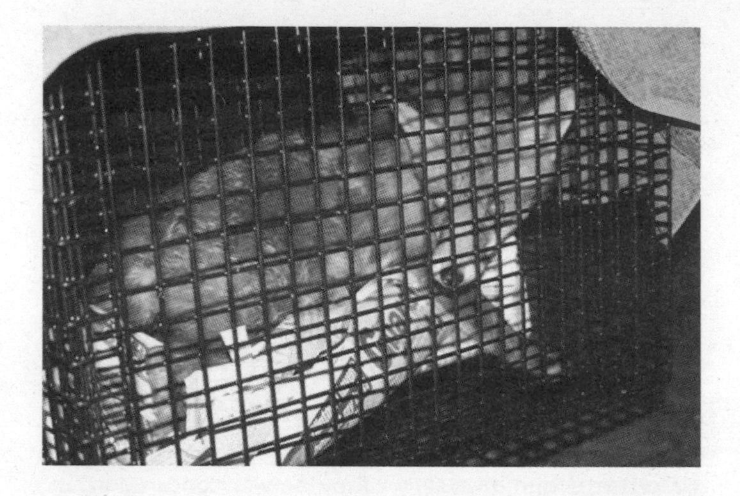

still so small in the carrying cage, she was brought back to us, and as the door was opened, she leapt with a single bound from the kindness of humans into the hazardous freedom of the wild.

She was very fortunate, and not only in the quality of her medical care. While she was being treated, those foxes who were so dominant over her continued to defend the group's communal territory. A vacancy does not take long to be noticed by transients. When the resident dies, reoccupation can occur within days – and a fox in a hospital is effectively dead, as far as fox society is concerned. Upon its resurrection, there may be trouble. Recent data from Dr Dawn Scott's research in Brighton suggest that re-released foxes wander much more widely than non-rescued individuals. Unfortunately, it is impossible to know whether the foxes in this study were transients before their accidents, but the risk of leaving a rescued fox homeless is clear.

Removing a fox from the wild may impact its life well beyond the few weeks that it spends in hospital. It should only be done when the fox's life or welfare are at serious risk, and no other method is viable. The preference to treat foxes in their natural environment has led some to promote unconventional remedies, especially homeopathy.

Psorinum 30c has become a major feature of many charities' anti-mange efforts. It is much easier to obtain than Ivermectin because a vet does not need to prescribe it. Unfortunately, that is because homeopathic treatments are medically inert; they are a discredited invention from the eighteenth century that has become a multi-million-pound industry.

I stir up this hornets' nest reluctantly and only because of my concern for fox welfare: I have seen that feeding homeopathy to a suffering fox delays genuine treatment and puts it at risk of death. To be clear, I have no problem with new or unusual techniques, provided they are supported by unbiased evidence. There is a $1,000,000 prize on offer in the USA for anyone who proves that homeopathy has medical value. It has not been claimed. This isn't the place to reprise Dr Ben Goldacre, although I thoroughly recommend his *Bad Science* book, but in summary: homeopathy doesn't work. It simply cannot work, because it conflicts violently with what we have since learned about biology, physics and chemistry.

Our eyes can deceive us, which is why good medicine is based on careful trials rather than anecdotes. *Correlation is not causation,* runs one of science's favourite mantras. It is not enough to say that A happened after B – you need to prove beyond reasonable doubt that one caused the other. In a fox with mange, there are many invisible factors that complicate the picture.

Once mange progresses beyond the early stages, many foxes suffer a terminal deterioration of their symptoms. But in others, the condition reverses. The bare blotchy skin patches and lesions remain, but the mites die, and the fox gradually heals.

Why this is, nobody knows. It could be that some foxes are less genetically susceptible. General health, stress levels, and the virulence of the mites themselves may also determine the outcome. In any group of infected foxes, some will live. I fear that homeopathy is taking credit for natural survival, an effect enhanced because its proponents encourage its use only for mild cases. These will almost certainly include foxes that are already on the mend.

It is possible that homeopathy can have an unintended benefit; by putting out regular food with the dose, a home owner may support the fox's general health and assist its fight against the disease.

Foxes face many hazards – natural and anthropogenic – in their daily lives. It is not surprising that so many do not see their first birthday, let alone their fifth or tenth. Longevity in the wild varies sharply between populations; it is not uncommon to see individuals above five years of age in my area, but they are unlikely to survive more than 18 months in London. The oldest wild foxes on record are a vixen shot in the Netherlands in her thirteenth year, and a Japanese fox who was aged to fifteen years by the cementum deposits in her teeth.

It is often a human hand at the end. Our cars, guns, garden pesticides and human-derived diseases do not put the fox at risk of extinction, but they have warped its world, creating a population skewed to juvenile level, and introducing forms of suffering unknown in the wildwood. Yet many people offer them sanctuary and healing as well. We are indeed a mixed blessing as neighbours.

10

Predators Among Us

You can change friends, but not neighbours.

ATAL BIHARI VAJPAYEE,
former prime minister of India

A FOX IS A fixed point of reference. Compare it to a billboard on a train station that turns the heads of a hundred commuters, each of whom will silently superimpose their own taste and needs over whatever is being advertised. How we colour the distance between us, is a product of us. The fox does not change.

'I THINK FOXES ARE EVIL.'

My colleague was distressed at a childhood memory of an attack on her pet rabbit by a fox. I had every sympathy for her. Losing a companion animal is painful beyond words. But 'evil' implies something more: that the fox can be expected to understand human moral judgement and is ethically bound to abide by it.

Is this reasonable?

There was a time when animals were tried in human courts. Medieval France began it by famously executing a pig for infanticide at Fontenay-aux-Roses; until the eighteenth century, many animals perished violently after being convicted of a crime. Most people today do not struggle with the idea that bringing animals to the dock is fundamentally flawed. Yet common speech often lets slip that our expectations of animal morality are more primitive than our legal system.

Nowhere is this more emotive than when a fox has destroyed multiple chickens or pheasants. Surplus killing – taking more prey than is immediately needed – is widespread in wild carnivores. Wolves sometimes kill dozens of elk or caribou. Lions kill hoofed mammals en masse; it happens today in Africa, and it was recorded thousands of years ago by the Old Testament prophet Micah. Grizzlies kill surplus salmon in unfathomable numbers on Canada's west coast every autumn, eating only the most nutritious parts and leaving the rest to rot.

Wild carnivores do not have moral culpability. Citing the empty coop as evidence of evil is philosophically void. Yes, there are solid factual reasons why the behaviour occurs – much the same, in fact, as your reasons for 'surplus buying' when you visit Tesco, and you would probably be surprised to be called evil for purchasing a week's supply of bread. Foxes, too, gather surplus food, for that is the instinct born of an uncertain world. They will try to bury the leftovers for later use. However, it is a very fine line between stating this important scientific fact and implying that the explanation means foxes are really keeping to human standards of morality after all. They are not, and we have no business promising ourselves that they will.

Foxes operate by instincts honed by the wildwood, not by modern human law. They cannot read 'no entry' signs. They do not have the mental or moral capacity to understand that wild rabbits are fair game while ones in a garden are off limits. They certainly cannot comprehend that part of their territory is forbidden because it is planted with exotic flowers.

If we are to live in peace, on terms that are genuinely compassionate for both rabbit owner and fox, we need to consider animal behaviour in a more realistic and dispassionate way. Neither the politician raging that foxes are bloodthirsty vermin nor the animal rights activist protesting that they never cause any trouble helps anyone.

We have built our world over the wildwood, and yet seem perpetually surprised that foxes are in it; it is a whimsy of the human race that we love wild animals that collapse towards extinction in our presence, and resent those that survive us. I have seen this same chill towards urban coyotes in Canada, langurs in Singapore, and magpies nearly everywhere. Wildlife that endures civilisation is seldom loved for it.

Or is it?

SOOTY IS A FOX in my village from a family that I call the Gatekeeper Group. I know her well, for she is not a difficult vixen to find. She trots down the roads at any old hour, irrespective of cars, dog-walkers and mothers on the school run wandering by. She ignores my leashed dog at five metres. She leaves eggshells scattered on roadsides and in flowerbeds – the eggs come from a local resident who appreciates her company.

If ever there was doubt that foxes can make people smile, Sooty settles the question. She is a friend – there is no other word – for a well-respected local lady who quite literally calls her by name. Watching this wild vixen trot down the road when summoned is a remarkable thing. She brings happiness to many, not through circus tricks but merely through the sheer fact of being alive. She visits multiple houses on at least three streets, and has won much laughter through carrying gardening gloves as well as planting the infamous eggs.

Sooty.

Yet the fox simply is; it exists as a native thread of Britain's natural ecosystems and thus has an intrinsic value whatever the current sway of political thought. I, like many others, enjoy watching them and my life would be far duller without them; but, in a way, that is just as irrelevant as the fact that other people despise them. The fox's ecological worth does not change in the turbulent seas of human opinion.

Nevertheless, there are several reasons why it is extremely important to understand how people perceive foxes. Negative beliefs frequently trickle upwards to councils and government and result in dismal, unscientific wildlife laws. Additionally, studying the human response to foxes provides a powerful insight into ourselves – fear of a 4.5 kg (10 lb) wild animal is credible evidence of how modern Britain is becoming increasingly disconnected from the natural world. Lastly, nobody can devise sustainable solutions to human–wildlife conflict unless they have a clear grip on the facts.

So what do people actually think? How much damage do foxes actually cause, and what can fox advocates do to reduce it?

In the autumn of 2015, I considered how to tackle this question. Gathering genuinely objective data on human attitudes towards foxes is not as simple as it sounds. Ask the question online, and you are likely to be swamped by one outspoken faction or another – or possibly both, and your post will burn red hot in a flame war. Post questionnaires through people's letterboxes, and only those interested enough in the topic will respond; this is also likely to lead to an over-representation of strong, atypical views. Standing in the street with a questionnaire alienates commuters and anyone else awkward about communicating with strangers – and as readers of the wryly observant Twitter feed *Very British Problems* can testify, that is no small percentage.

Capturing the quiet voices of those ordinary people who share southern Britain's cities and villages with foxes requires careful planning. Together with Marc Baldwin from wildlifeonline.me.uk, I commissioned a professional opinion poll company to approach over 2,000 people. Critically, the respondents were a general snapshot of British society – they hadn't been pre-selected because of their interest in the fox debate. Young, old, city-dweller or rural born and bred: they were given the chance to state in their own words how they viewed the foxes in their midst. The results were astonishing.

Three-quarters of respondents claimed to either like having foxes in their neighbourhoods, to feel no strong opinions about them, or to believe – probably incorrectly – that foxes were absent in their area. The other 25 per cent expressed outright disapproval. But what leapt jarringly forth was the vast gap between London and people in the largely rural Home Counties. The latter returned a fox disliking percentage of 19 per cent. In London, it was 33 per cent.

Adult male Londoners were the least fox-sympathetic of all respondents; London women, however, were the most fearful of an actual fox attack. Meanwhile, women in the Home Counties scored the highest level of liking foxes.

What motivates people to love or loathe foxes?

They look very threatening and I always worry they will attack my children. They are dirty (rodents) and they go through the bins. I would like them to be hunted and killed. They are RODENTS, explained one respondent. While the biological error appears to be an uncommon belief – most people surely recognise foxes as members of the dog family – this comment does highlight several 'disliking' factors. The top three complaints were raiding dustbins, making noise, and leaving scat. Worries about the safety of pets and children also scored fairly high. Surprisingly few respondents suggested any concern about garden damage or zoonotic diseases.

The conservation biologist wanting to reduce human–wildlife conflict first needs to know what conflict genuinely exists. It is a sobering reality that some of the most explosive complaints about wild animals bear little resemblance to what wildlife actually does. There are places where the old narrative of 'solve the problem and hatred will vanish' still holds true; the wildlife conflict teams helping villagers and lions to safely coexist in rural Gujarat are an outstanding example. But such cases are vastly outnumbered by situations where the anger is built on exaggerations, misconceptions and – all too often – a proxy of some local human politics. Cold hard science can help when a shepherd needs assistance protecting his goats from hyenas; faced with those who hate the same hyenas because of an ideological grudge and subconscious brainwashing from Disney cartoons, it falls painfully short.

With that in mind, we have to divide alleged fox problems between those that cause tangible harm and those that exist only as political constructions. The first are obvious: physical attacks on children, pets and livestock, damage to property, and transmission of disease. These are all matters where scientists are primed to intervene.

How significant are these real problems?

Obviously, statistics mean little to an individual rabbit owner who has lost their pet. There is no doubt that foxes will take small mammals and domestic birds if given the opportunity, just as they respond to opportunities to catch mice and woodpigeons in the wild. Prevention is the only solution. Rabbit advocacy groups advise supervising bunnies that are loose in the garden; pens should be weld mesh rather than chicken wire. Alternatively, why not turn your rabbit into a house pet? My family used to own an enormous lop-eared albino who regularly hopped around the house. Rabbits are intelligent and highly social. Foxes or no foxes, they need more from life than to be left in a hutch all day.

With hens, it is as well to remember that a fox can easily clear a six-foot fence. They are more inclined to dig underneath obstacles however; an L-shaped fence base (also known as a no-dig skirt) can be effective. Electric fencing is also often used but it is important to keep it operational at all times. For larger flocks, leaving a hen-friendly dog, pig, or – should you feel exotic – a llama in the exercise area may be an option.

The cat question is a much more vexed matter, and recent media reports about the so-called M25 Cat Killer have amplified concerns. This sad saga began when an animal rights group made the explosive claim that a human criminal was systematically stabbing and dismembering beloved pets, mostly in the London area. Like many theories that court our worst nightmares, it ignited on social media with a passion too urgent to allow objective consideration of the evidence. Not only were cats being tortured, we were told, but also rabbits and foxes – and maybe a human victim would be next.

Yet there was no killer. In September 2018, after three years and unknown quantities of taxpayers' money, the official investigation discovered that – foxes scavenge dead animals. The cat killer was a motor car. The mutilations? Innocent post-death scavenging; and yes, carnassial teeth will leave marks that resemble knife wounds, not least because they are designed to provide wild carnivores with the function of a knife. *Unbelievable,* social media exclaimed; but anyone who cares for either cats or foxes should believe the cold bland facts.

To be absolutely clear, the police did not claim that foxes are predating cats. What they did say is wholly consistent with the forensic evidence, known animal behaviour and common sense: cats are often hit by cars and then partly consumed by wild scavengers. Rabbits are likely to be the victims of actual predation. In either case, the actions of the feeding fox will often lead to the dismemberment of the body. It is, of course, very distressing for the owner to find their pet in such a state. What about the supposedly stabbed foxes? Nature is not squeamish about cannibalism. I once found an adult fox carcass in a wood which showed evidence of post-death mammalian scavenging over some days, and while there is a small possibility that a badger was at work, my suspicions are mostly with the local fox group. This is not uncommon in the wild, from bears to spiders. I have even seen a species of Canadian squirrel eating its traffic-killed kin.

It is dreadful that many grieving pet owners were caused unnecessary anguish by the claim that their animal had met a grisly end via a human monster. It is also unfortunate that the slightly garbled reporting poured fresh doubt into the sensitive topic of how foxes and cats really interact. Reality, however, is rather calmer. Even assuming that DEFRA's estimate of 430,000 British foxes is correct, cats outnumber them by nearly 19 to 1. On any given night, a fox will encounter many cats, and the normal outcome is for the two animals to ignore each other. I caught this on a trail camera in my local town – unseen by any person, a ginger tabby stands aloof as a fair-sized fox passes it and continues down its route. A healthy adult cat is a formidable beast with teeth and claws that a fox can only dream of – there are reasons why rescuers capturing feral cats are extremely cautious. Some cats are fiercely territorial and make a hobby of driving foxes out of their garden. That said, kittens or very elderly cats are at more risk from all the dangers of the outside world and should not be allowed to roam unsupervised.

What about garden damage? To some extent, it is a matter of tolerance. Digging in the lawn annoys some, but others accept it. Using bone meal or dried blood as fertiliser increases the likelihood of digging in flowerbeds. Common sense may also resolve other complaints; foxes do not raid dustbin bags that have been stored in secure wheelie-bins. Where a gardener seeks minimal contact with foxes, the commercial deterrent Scoot may assist. As previously discussed, this non-toxic spray prevents foxes adequately marking their territory, which is highly off-putting to them. Attempting to fence foxes out is likely to fail, and may harm other species such as hedgehogs.

REAL CHALLENGES need real responses. But the second category of fox 'disliking' is ideological. Some people are genuinely offended by the idea of a fox walking down a street even when it has done them no harm whatsoever. This is problematic. Science cannot slay dragons that do not exist; it cannot solve problems that are not occurring. A fox is not a problem neighbour for the mere fact of being alive. This type of fictitious but brutally bitter 'wildlife conflict' is certainly not unique to Britain, but it is not to our credit that it has become entrenched here. Unfortunately, stating scientific facts to a person who dislikes animals on ideological grounds seldom improves anything. Building tolerance takes education, patience and time.

But these people who harbour hostility towards foxes are a small minority. Just as illuminating were the respondents who explained why they *like* having foxes in their midst. *It's nice to feel that nature is still around us and that we haven't destroyed everything,* one respondent wrote. This sentiment was widely echoed. Foxes are indeed valued for the connection that they provide with the raw, non-human universe. They are a small spark of wildness amid our chaotic lives.

But perhaps the most fascinating result was that the 'good' qualities cited in foxes overwhelmingly influenced a person's overall opinion. Those who valued foxes for their utilitarian purposes – catching rats, for example – still tended to view them rather poorly. A high level of fox appreciation was strongly linked with accepting that the fox has intrinsic value.

This is the key message for wildlife educators. A person who has suffered harm from a fox but places a high moral value on coexisting with wildlife is more likely to be tolerant than a person who never experienced depredation but harbours resentments for reasons far beyond science. Trying to justify the fox's existence with facts about its usefulness may, in the long run, be less effective than simply encouraging an ethic of respect and coexistence.

Foxes are called a complicated species. So, undoubtedly, are we.

11

When the Fur Flies

> It was our War of Win-Nothing . . . nobody won, Madam, not the people and not the tigers. It was the most futile of wars.
>
> *The Singing Mountain*

I KNOW A MOUNTAIN where wilderness is still truly wild, far away on the eastern slopes of Canada's Rockies. I have walked there in autumn when trembling aspens paint the ridges with golden leaves, and seen springtime windstorms ripple vast wildflower meadows like waves in the sea.

One morning, I was wandering the streets of the little village under that mountain, and happened across a store tended by an old lady agleam with the health that only comes from exposure to pure cold wilderness air.

Pumas were our topic, because she had seen their massive teardrop-shaped footprints right here among the houses. She had also known bigger neighbours.

'You have a grizzly in your yard?'

She smiled, wrinkles carved by a thousand Rockies storms dancing about her eyes. 'I keep the grandkids indoors when the bear's here with her cubs, but they love watching them. It's good. It's the bears' land.'

Grizzly bears can and do kill people, but she had chosen to humanely coexist.

The date was 6 June 2010. Looking back now, the irony is not lost on me. That very day on a street in west London, a highly habituated fox trotted past a three-storey townhouse occupied by a family and their two baby twin girls. Within twenty-four hours, a storm more hellish than anything curdled by Alberta's mountains would ignite.

I know that many people are afraid. A wild animal attack is uniquely traumatic and echoes in our subconscious long after it has slipped from the press. I have observed that anxiety in the shadow of the world's greatest wildernesses, and I have seen it under London's skyscrapers too. It often bubbles over into the irrational, from tribal killings of harmless Andean mountain cats in Bolivia to grossly exaggerated, lawyer-driven black bear advice in North American national parks; but let there be no doubt: wildlife attacks do occur. Our challenge as a species is to respond to that reality in a humane and proportionate fashion.

Exactly what led a fox to enter the Koupparis family's property in Stoke Newington that June and repeatedly bite both twin girls may never be known. If foxes bit children every week, a pattern of risk factors might be discernible; faced with a behaviour that is lightning-strike rare, it is difficult to form conclusions. I do not think that speculation and finger-pointing benefit anyone, least of all the human victims.

The media did not share this view. For a moment as I stepped off the plane from Alberta, I wondered what had happened to my country. Newspapers, television screens, the internet – they burned red, red hot: raging at foxes, raging at a previously anonymous family for being bitten by foxes, and reserving a particularly corrosive kind of rage for anyone who complicated the narrative by not being either an extremist fox-hater or an extremist fox-attack denier.

While scientists wish to reduce fear, some journalists strive to increase it. Barrels of ink and forests-worth of trees were sacrificed in bizarre tabloid photographs of foxes eating biscuits – carefully edited to give the idea that their jaws were open in a menacing snarl. Deprived for so long of the dangerous wildlife that haunts the psyche in Australia and the USA, the papers pounced on the story with an unsavoury relish.

Wild animals are easier prey for newspapers than celebrities: they cannot sue. If they could, at least one tabloid would go bankrupt tomorrow. But while some wildlife advocates fought the anti-fox fever using the official channels such as the Press Complaints Commission, others turned their fire on the London family at the centre of the storm.

It was not a flame war. It was too one-sided. Social media pages became filled with brutal anger. 'Report [the mother] to Social Services for child neglect!' shouted one Facebook post, apparently started on a well-known fox advocacy page and shared to kingdom come by its members. The lack of empathy for the twins' parents was chilling; no accusation seemed too hurtful provided it directed attention away from foxes and recast humans – however innocent – as the villains.

To this day, some fox campaigners do not accept that the swirling rage approach might be flawed. This has not gone unnoticed. When the pro-hunting Game Conservation Trust sent a representative to my university, he used the abuse received by the Koupparis family to convince students that fox advocates are inhumane nutcases.

Benefit to foxes, then?

That would be zero.

What went so catastrophically wrong?

'FEAR IS THE PARENT OF CRUELTY,' writes James Anthony Froude. But the deepest fears spring from love – for the safety of those close to us. It prompts the questions that London mothers ask about foxes. Likewise, the extreme knee-jerk reaction of some fox admirers to the accusation of biting may be born out of fear that acknowledging any 'misconduct' will open the door to a cull. But 'normal' people who simply have a few doubts about foxes in their garden aren't generally comforted by angry denials.

Of course, many wildlife advocates and university professors in England do deal with situations like this with sensitivity, but outright disavowal is more common among the public than in animal bite cases in North America. This may be a question of familiarity. While the risk from bears, wolves and coyotes is far less than is generally perceived, it is much harder to put forward a never-will-bite mythology when, tragically, fatalities have occurred. Wildlife advocates in the USA and Canada know that they are caring for animals that can be lethal. It doesn't stop them caring.

Coyotes, distant relatives of the fox, are a classic case. Perhaps the most persecuted large mammal on the planet, these small, highly intelligent wolves are mocked in common culture and subjected to shockingly diverse cruelty every year, everywhere. Nevertheless, they have brave advocates who encourage urban people to live safely and peacefully with the coyotes in their midst.

One of the leading lights is Emily Mitchell, who founded the Taylor Mitchell Legacy Trust after her beautiful daughter was killed by two coyotes in Nova Scotia. While the province's government responded with a coyote bounty in defiance of all scientific advice, Mrs Mitchell believed that her daughter deserved a better memorial, and now campaigns tirelessly for the safe appreciation of wildlife. She is a humbling and inspirational woman.

Britain is not so familiar with threatening animals, and perhaps the biggest cause of the fox bite furore was its sheer novelty. Humanity, like foxes, gives new experiences special attention. Rare types of death dominate our press. We all know that heart attacks and motorbike accidents are much more common than terrorism or plane crashes, but that does not reflect the headlines – or, perhaps, the private fears of the average person boarding a flight.

Not only are fox bites extremely rare, but they also touch that primeval sore spot. Perhaps our fixation with animal attacks is written into our DNA, dating from a hazy prehistory in which humans were regularly predated by great cats. The media delights in negative animal stories for a reason: they sell.

Perhaps, also, it is a twisted acknowledgement of our broken relationship with nature. Once upon a time, children knew that they walked on a living planet by harvesting mushrooms, climbing hollow trees, or rubbing stinging nettle wounds with dock leaves. To the person in a nineteenth-floor Croydon apartment, the jolt of a headline about false widow spiders is a reminder that, even in modern Britain, the natural world and humanity do interact.

Be that as it may: how can we keep the response to urban foxes proportionate, and what factors increase the odds of negative encounters?

In the aftermath of the attack on the Koupparis twins, many fingers pointed at the practice of deliberately feeding foxes. It is beyond doubt that providing handouts to foxes is encouraging higher than normal population densities, with individuals cramped in greatly reduced territories and travelling shorter distances each day.

It is also indisputable that feeding is a direct cause of coyote, dingo and wolf bites in other parts of the world. Usually, it happens like this: a coyote trotting past a campground in, say, Yellowstone is thrown a sandwich by an intrigued visitor. As an intelligent animal able to respond to changes in its environment, the coyote remembers this experience and returns to the campground. One day, someone jerks an arm back too fast, or emits some

Coyote begging a tourist for food in Yellowstone, USA.

signal which the coyote misinterprets as aggressive – and it bites.

The sad reality is that it is seldom the person who starts the cycle who suffers. It is the War of Win-Nothing: everyone loses. Including people, for being bitten is terrifying. The coyotes also lose. They are invariably shot.

Can we extrapolate this to fox encounters?

Not perfectly; there are certainly lessons to be learned, but foxes are only distantly related to the genus *Canis,* which contains dogs and coyotes. They are far less aggressive in their intraspecific social lives; while wolves often kill each other, quarrels between adult foxes are less frequently fatal. As lone hunters, they have limited 'rules' for soliciting food from each other. They are also physically weak and almost never predate anything larger than themselves, unlike the *Canis* species, some of which regularly kill moose and bison. Except for very young children, all fox bites are defensive or inquisitive, not predatory. In other words, foxes aren't actively looking for opportunities to bite us.

Moreover, the numbers do not lie. Relatively few coyotes are ever fed by people, yet it is common for these particular individuals to become biters. Foxes have been widely fed in the UK for decades, and despite ample opportunity, they remain almost entirely harmless. They usually wander past houses without going in. They see tens of thousands of children every day without approaching them. They trot past walkers without displaying any begging behaviour. There are sad exceptions, but the percentage of fed foxes – or any foxes – showing seriously undesirable behaviour is vanishingly small.

Fear of foxes is not simply supported by the cold, objective facts. There is no disaster here, no invasion of dragons. We have lived safely among them for generations. What we do have is a very slight risk, and, being human, we wish to empower ourselves by reducing it further.

At the same time, if we are ethically comfortable demanding that desperately poor developing world villages conserve lions, leopards and tigers, perhaps London should not bristle at humane coexistence with a far smaller and milder carnivoran.

Should we ban feeding?

The debate has been muddled. The risk is already so minute that a blanket ban on feeding cannot achieve any quantifiable results – almost-nothing to almost-nothing will not show up in any statistics. So many fed foxes do not bite anyone that it is impossible to state categorically that preventing feeding will improve safety.

Reducing the available food would cause local fox densities to decline, however, and other unpopular behaviour, such as damage to lawns and intrusion into houses, would probably decrease also. On the flip side, countless families would lose an activity which provides a link with the natural world. Feeding creates a bond. Human tradition across all cultures shares food in hospitality, and we have extended that to pets and local wildlife. The instinct is so strong that some people continue to provide sustenance to wild animals even when presented with clear evidence that it is doing them harm.

Feeding foxes is so ingrained in our culture that any ban would probably be unenforceable. And what message would such a law send? I would not wish for the UK to follow the ideals of British Columbia, where I have personally witnessed how intense wildlife phobias are harming both animals and people. Unbridled fear alienates us from nature, and motivates severe mistreatment of animals without increasing anyone's safety. I have seen this cycle. It is not pleasant.

Personally, I far prefer to watch foxes being wild – hunting in meadows, drinking from streams – but recognise that feeding them in gardens is extremely popular. There is nothing inherently wrong or dangerous when a fox is relaxed at the sight of a human. I've come across many 'bold' foxes in the vast Canadian wilderness too, where they are never fed at all. Most foxes that look tame are in fact fully aware that they can bolt to some secret hole under a fence if the person they are watching comes any closer. The ones that persistently beg, investigate living rooms, scratch at patio doors and closely follow people walking down the street are different. They are not born; they are made.

Begging is learned. That is not a fox problem – it is a human behaviour problem. If you do feed foxes, the golden rule is to remember the big picture. Do you live on a 1,000 acre farm? Then nothing you do is likely to turn your neighbours against foxes. Or do you have a 10 foot garden in central London? If so, does the young mother next door feel comfortable at being approached by a wild animal that has been taught that humans mean easy handouts? If she isn't, is it in the fox's interest to be encouraged into an area where it might be at risk, or at the very least cause tension?

Perhaps compromise should guide the fox watcher's code of conduct:

- Never hand feed or attempt to tame a fox.
- Never, ever, permit a fox to come indoors.
- Avoid feeding foxes right on your doorstep; use the back of the garden instead.
- Keep food quantities small; do not overfeed. Scatter raisins or dog biscuits if you wish to watch them.
- Avoid highly processed and otherwise unsuitable food such as cake and chocolate.
- Encourage others to enjoy watching foxes, but correct misconceptions politely – online or in real life.
- Don't deny that foxes can cause problems. If someone is worried about their children, lawn or livestock, refer them to humane deterrents like Scoot.
- Don't assume that all reported fox bites are a myth. In the unlikely event that it does later turn out to be false, you will still have helped promote humane coexistence by taking the high road.

And let us all learn to live together in peace somehow.

12

Tornado in a Cage

A robin redbreast in a cage puts all
Heaven in a rage.

WILLIAM BLAKE

SEPTEMBER SKIES ARE cold and quiet. Thin cirrus clouds appear daubed on pale blue as if with a painter's chisel, decorated with swallows returning to Africa. The chill heralds morning mist in Surrey's valleys, while squirrels tumble from orange-tinged trees as they gather conkers, and people and foxes taste the sweet juice of the blackberry crop. Autumn is British nature at its richest.

But one September, a piece of wildness was behind bars in my garden. All around her, trees forsook their leaves, mice buried sycamore seeds, and jays gathered acorns, but she was segregated, cut off from the rest of the British ecosystem by wood and wire. Nature is a tapestry where untold millions of living things interact. Alone in her pen, she made no more sense than a single splosh of paint removed from a watercolour.

Andrea was a rough, tough fox from Battersea; amber eyes so wild, brush so thin from mange. Her prison was her hospital. She was in a cage because she needed to be, having been discovered with mild mange in a brick-walled London garden, but no human child ever appreciated her doctors less. It was a privilege to care for her, but she was no pet. She was wildlife. More, she was a caged tornado, as intense as a hunting cat, as restless as a river.

Her accommodation was a rabbit hutch surrounded by straw. It was designed for her comfort, but whatever emotions of displacement and tedium were agitating within her, the pen paid the price. She began by stripping wood from the rabbit hutch, and shunted the entire heavy structure inwards at a diagonal angle. Having thus created a triangle-shaped gap, she curled up within it, eyes baleful yet very bright.

Over the following days, as her mange healed, she tore a hole in the hutch's roof. Even feeding her was challenging, for she destroyed every dish provided. Plastic trays were bitten in half, and china water bowls flipped upside down. I purchased a plate-sized bowl designed for a Great

Andrea.

Dane, but this small, lost carnivoran still managed to turn it over.

I did not pester her, but whenever I had to enter the pen to check on her, she retreated to the hutch, watching me through the hole that she had carved. She uttered no noise but questioned me with her eyes. For all their varied calls, a silent fox can be the most expressive of all.

Her *pièce de résistance* then came. She discovered one of my long-suffering trail cameras – it kept filming as she kicked straw over it, but that was the last experience it ever documented on this earth. She lifted the device from the pen wall, carried it to the far side of the enclosure and buried it in her bedding. It never worked again.

But I understood her meaning, for there is no keeping such an intelligent, sentient wild being in a pen without consequences, even when the captivity is in her best interest.

Of all the foxes that have been under my care, Andrea moved me the most. Her thin scarred muzzle, the brilliance of her intense amber eyes, the sense of knowing, as they say with wolves, that the creature under observation is looking back at you with conscious thought – she was tattered, but she still needed on the deepest level to be free.

She needed to be challenged by the daily quest for food, to learn her map and memorise routes, dens and experiences. She needed, too, to scream at rival vixens and bark for a mate, and to leap on voles, bury golf balls and listen to sounds so faint that humanity must use machines to detect them. Removing these challenges did not extinguish the energy built into her instincts to meet them.

Put simply, she was bored. The instinct, the impulse, the drive to be a fox was not separated from her by the walls that contained her.

After a few days, she was collected by other volunteers and deemed healthy enough to be released. Battersea is no Garden of Eden for foxes; the old power station may be patrolled by peregrine falcons these days, but the rows of low terraced houses, so often the hunting ground of drug dealers, offer fox habitat only in fragments. Yet it was her home, and where she was meant to be.

EVERYONE ON THE fox front line, whether in wildlife education or rescue, has been asked the question. The answer is: 'No, I do not recommend keeping foxes as pets. They do make awesome wildlife. Please appreciate them as that.'

Wildlife intrigues humanity. That is healthy and natural. Some express that curiosity through the sciences, and others through art, photography or simply watching. Most of us understand that there are two players in this game; while satisfying our need for contact with the natural world, we respect the boundaries that wild creatures would prefer us to keep.

There are others who interact with nature in less sympathetic ways, be that with a gun or the type of shockingly inconsiderate, disturbance-heavy wildlife watching common in places such as Yellowstone and Banff National Parks.

And there is another, often overlooked group who seek contact not by travelling into the wild world, but by bringing wildlife into their very human homes.

I saw one of those people build a cage out of a small, unkempt living room in Canada, furnished with an off-pink dirty carpet and a white sofa embroidered with flowers. Children's toys scattered over the floor, the unwatched television blaring out James Bond songs from the corner – I can still hear my co-rescuer gasp.

Eyes met ours: black, gentle and trusting. Their frame was spotted fur supported by fragile thin legs. It was a deer fawn, in a house, involuntarily watching MTV.

The master of that house had stumbled across the fawn while walking in the moss-choked temperate rainforest that mantles Vancouver Island from rocky shore to mountain ridge. The fawn, he said with no emotion, was supposed to be a pet for his six children, but three days on they had all become bored with it. Worse, the pink plastic baby's bottle used to feed it contained cow's milk, all but guaranteeing potentially fatal diarrhoea.

One child extended a thin hand and patted the fawn as we carried it away – the last vestige of dying novelty – but there was no ceremony to our leaving. Past illegally dumped garbage that had not long ago lured a bear into town, into a 4x4 car and down the gritty hill towards the sea – still the deer came with us, a wild creature afloat on the dilapidated edge of the human world.

I do not care to recall the next two days. In the roughest brush strokes, imagine a forest so damp and dark that the vehicle track feels like a corridor draped with thick green curtains, behind which tower sheer grey mountains eternally drenched in rain-fed waterfalls. Imagine searching up and down, up and down 60 km (27 miles) of that broken road for one deer mother, while starless night falls over you and the fawn's cries of hunger merge into a migraine from hell.

One choice, one human action, one belief that wildlife can be magically transmuted into a pet on an impulsive whim changed that fawn's life, breaking not only the law but also an animal's connection with the ecosystem that had fed it, rained upon it, and perhaps would have one day claimed it as prey for a wolf or bear.

It seems almost incidental in comparison, but that one action also plunged multiple humans into days of utter chaos. After a grim number of phone calls and emails, I learned that a wildlife rescue group was willing to, well, rescue me. Sadly, they were on the island's other coast, a full six hours' drive away.

We met halfway, at a gas station in a wilderness village. Forest and cloud-wreathed mountains loomed over us as the fawn changed hands. I do not know if it survived because I never found the will to ask.

BE IT DEER, foxes or pandas, much wildlife is within walls built by humans these days. Outside of zoos, captive foxes in Britain can be divided into three categories. Firstly, there are wild foxes in the care of charities and their volunteers; such care is usually temporary, as in Andrea's case, but may be lifelong if the fox's health makes survival in the wild unlikely. Brain damage from toxoplasmosis, for example, can create a fox which shows little fear of anything and is much safer in a pen.

Secondly, there are silver foxes kept by exotic animal enthusiasts. Descendants of the trade in fur farming, they have North American origins, and it is noteworthy that some authorities consider the American red fox to be a separate species to that indigenous to Britain. These foxes are not fully domesticated – they are considered to be farm animals under EU law – but display some physical and behavioural characteristics that differ from 'made by nature' wild foxes. Their coat colours are far more variable, for example. Caring for such an animal is a complex responsibility that silver fox advocates insist should never be commenced lightly.

Finally, there are those wild, British red foxes who find themselves on the leash.

Often it is well meaning. But finding a helpless cub is not an invitation to take it back to your living room. It will end in tears, quite possibly literal human ones; the RSPCA and Scottish SPCA readily seize such 'rescued' foxes and will hand them over to experienced charities who will care for them, not as pets but as the wildlife they are.

Nobody can pick up a wild canid cub in a suburban hedgerow and replicate 36,000 years of dog evolution with a few hopeful words. No amount of hand-rearing, promotional photoshoots with the tabloid press, or mealtimes with Bonios can alter the reality that a fox is intrinsically a fox. They defecate in their food bowls. They solicit attention through nipping. They have an odour – arguably more pleasant than pub vomit or burned toast – but it is strong enough that even the slightest dash of urine can be detected by people from several metres away.

Wild orphans need expert care, not just in terms of food and medication, but more critically to bond with their own kind. A responsible wildlife hospital will always raise foxes with other foxes. Any photographs showing a fox cub being reared by a dog, cat or other unlikely mother are highly suspect in the welfare stakes.

The interests of an orphaned or injured fox are best served by ringing immediately for expert assistance. The helpwildlife.co.uk website will enable you to locate a nearby centre. I strongly recommend keeping your local wildlife hospital's number on your mobile phone.

Native foxes that are bred in captivity in the UK will only show a thin veil of domestication; the same irrepressible vulpine instincts lurk beneath the collar. They do not turn into cute orange dogs just because they are called 'fur babies'. It is also worth noting that some controversial facilities, such as the notorious Miyagi Zao Fox

Village in Japan, exploit the 'cuteness' angle to encourage visitors to admire foxes in conditions that are far below acceptable welfare standards. We live in an age obsessed with anthropomorphic cuteness, but often at a sad cost to the animal itself. Basic rule of thumb: if that sweet photo on Facebook or in the *Metro* looks too cute to be true, it probably is.

Perhaps the fox on a leash is the fault of the domestic dog. They are so ubiquitous, so happy to share our lives – their brains light up with oxytocin even as they stare into our eyes – that we forget how special they are. Dogs are unique. They are practically symbiotic with human beings; we have evolved together for many thousands of generations, far longer than any other domesticated species. The image of some chest-thumping cave-dweller stealing wolf pups from a den and brutally subduing them, forcing them into a new, civilised way of life, is not supported by recent studies, which suggest that wolves – or very early dogs – actually chose us. They scavenged our refuse, perhaps shared in our hunts. They changed in our company, assuming a less aggressive social life than the ancestral wolf, their internal organs altering to adapt to a less meaty diet.

Dogs have also changed us, playing a pivotal role in the human migration into the Americas, and of course now assist every corner of our Western civilisation, from capturing criminals to comforting sick children.

This is not what usually happens when people and wildlife encounter one another. There are approximately 5,500 mammal species existing on Earth today, and only one has partnered with humanity to this level of mutual trust. Later, horses and cats also entered our emotions; we certainly use and abuse other species, converting jungle fowl into battery-farmed chickens, and gracious mouflon into sheep suffering from foot rot, but dogs are more than livestock to us, at least in the West. For all their important work, for the most part we simply consider them to be friends.

WHEN I DISCUSS pet foxes online, the inevitable rebound is: 'But what about the Russian foxes?'

It is true that an experiment was started in Soviet Russia by Dmitry Konstantinovich Belyayev in 1959 that subjected caged foxes to an intense programme of breeding selection. The 'friendliest' were selected to pass on their genes in the hope of achieving a super-speed domestication. After some generations, the most tolerant foxes had begun to develop other curious traits such as curling tails and patchy colouration.

But a fox is not an inanimate object. For all its fame, this project should ring serious ethical alarm bells, including the fate of the discarded foxes (frequently slaughtered for fur) and the conditions in which the animals are maintained. This research was undertaken solely for human curiosity; the argument that it explains genuine, prehistoric domestication of dogs is hard to swallow, for the conditions are wildly different. Research that carries significant welfare costs has to be weighed against conservation or scientific gains. It is hard to see what foxes themselves have gained from this project.

Even 'tamed' wildlife can pose a safety risk; in the United States, where 'pet' tigers are widespread, deaths have occurred. It is also true that exotic pets rarely live out their natural lifespans; once the novelty wears off, they are either put to sleep or confined to a pen, never to exercise their natural instincts to wander and interact with the rest of the ecological world. There are also health concerns. Captive prairie dogs have transmitted monkeypox to their owners, for example, and few vets are trained in medicating exotic pets.

Sadly, it seems that none of this matters to some of the wild pet market clientele. I have seen advertisements for exotic cats that tempted buyers with 'Think of the looks you will get!' This is not really a pet – it is a status symbol, or sometimes part of a collection that can be traded like stamps. Anyone who has had a friendship with a dog will find this hard to understand.

So, if anyone wants a true, heartfelt bond with an

animal that will happily adore you for its lifetime, accept training, and accompany you on walks and adventures, you can do something amazing: acquire a dog. If, however, you want an animal for boasting about – well, with respect, it's probably best you don't keep any pets at all.

But there is another question, one I have pondered while my dog plays staring contests with the foxes that peep out from shadowed driveways in my village at night.

None of us will be here in 36,000 years to see what has become of the human–fox relationship. Yet the similarities with ancient dogs are tantalising. Like proto-dogs, they scavenge our food. Like proto-dogs, they are adapting their behaviour to orbit around us. They lack the dog's great ability to read human body language – but then, dogs outperform everything else at that, even wolves and chimpanzees. And foxes do interpret some human behaviour; they know when an outside light means a house is occupied, and recognise plastic tubs used to feed them, or come running at the sound of an open patio door.

At what point does the fox stop being built by the wild, and start to redesign itself on us?

Perhaps, in some far, far future, there will be a creature curled up on our doorsteps that is as different from the wild fox as the dog is from the wolf. While foxes lack the folds in the pre-frontal cortex that are associated with highly complex social lives in wolves and their kin, they are still more group-orientated than cats, which have settled in our company following 9,000 years of domestication.

But for the here and now, they are still wildlife.

And I hope that some foxes at least will always be free.

EPILOGUE

Fame and Foxes

THE MAGIC OF FOXES is their behavioural plasticity; even the most mundane sighting can end with a surprising twist.

Romeo

There is a glass spire near London Bridge railway station that looks like it has tumbled to earth out of the fantasy worlds in the *Myst* series of adventure video games. At over 300 metres (1,000 feet), the Shard is Britain's tallest skyscraper, and for an equally giddy expense offers visitors an eagle's view of the sprawling metropolis and the fragments of surviving countryside beyond.

One of the Shard's most famous visitors did not a pay a penny. In 2011, while the edifice was still under construction, a young male fox entered the skyscraper and climbed. And climbed.

He reached the 72nd floor, where bemused builders fed him scraps until a wildlife rescue group arranged for his capture. After a check-up, he was released safely on terra firma.

Every time I pass this soaring glassy building, I wonder how he managed it, but the quiet question of *why* proves even more unanswerable. He was christened Romeo by his rescuers and has since been immortalised in charming illustrated children's stories that are sold in the Shard's gift shop.

Imelda

In *Dora the Explorer,* Sneaker the fox is a thief. It is hard to dispute this presentation; they readily collect and cache almost any item that can be carried. Underwear on washing lines, dog toys and footballs are taken by some; golf balls are not infrequently stolen from courses and buried in flowerpots. One transient dogfox visiting my garden walked off with an entire birdfeeder filled with peanuts, burying it in my neighbour's flowerbed. Even more startling was the fox who trotted past one of my trail cameras with a toy dinosaur in its jaws.

But few foxes will ever reach the magnitude of kleptomania achieved by Imelda. A vixen in the German town of Föhren, she gathered shoes to a fantastically obsessive level. For over a year, steel-capped workmen's boots, wellingtons, and slippers all vanished from doorsteps. When a forestry worker stumbled upon her den, an astonishing 86 shoes were stashed around it. Another 32 were found nearby in a quarry. The count of the town duly had laid them out in his palace for their owners to collect – and put out a gentle advisory to keep footwear indoors at night.

Nevertheless, enough prize footwear remained within vulpine reach for Imelda's quest to continue. By the end of 2009, she had allegedly stolen 250 shoes. The following year, she managed to reclaim one of the pairs that had been rescued from her den. The townsfolk responded to their obsessive fox with good humour.

Commuter club

Foxes travel. For millennia, they have crisscrossed the northern hemisphere on their own feet – but urban humanity has widened their options. Astonished London tourists learned as much in November 2017, when they noticed a fox apparently sightseeing on the upper deck of a tour bus. After being treated to some of the world's most famous landmarks, the fox was escorted off by the RSPCA, who released it back in its territory by the bus depot.

Not all travelling foxes have needed human help to disembark, however. Another London fox made headlines for jumping on board an ordinary commuter bus during rush hour and getting off at the Imperial War Museum. Yet another was observed taking the escalator at Walthamstow Central tube station. This is not only a British phenomenon, either; a fox was recently photographed riding a Melbourne tram, while another made Canada smile through falling asleep on the seat of an Ottawa bus.

Social stars

As we have seen, foxes have a mesmerising, almost cat-like hold over social media. It is true that no YouTube clip featuring a flesh-and-blood vulpine is yet to match the 800 million views achieved by Ylvis in their official music video for *What Does the Fox Say,* but nonetheless there have been plenty of unexpected stars. Foxes playing on back garden trampolines are a particular favourite, while movies showing a hunting fox leaping headfirst into a snowpack often go viral. Other foxes unwittingly enjoying a few minutes of fame include one that spent a morning playing like a puppy in an urban hedge (nearly 7 million views on Facebook to date) and another that danced in the snow with a stolen dog toy.

Foxes have also carved a place in the blogsphere. Apart from my own site, they star on the long-running *Every-*

thing is Permuted blog by South Downs-based Words (Paul Cecil). Twitter is not immune either; popular feeds include *The Hourly Fox,* alongside many nature photographers who post pictures of foxes when the opportunity arises. It seems likely that the fox's special formula of being not only intriguing and beautiful, but also easily visible for millions of people in town and country, will keep it at the forefront of social media for many years to come.

Brandy's playmate

Back in the days when dogs could roam free, a lady in my village listened to a fox barking on her front lawn. As was done in those times, she put Brandy, her terrier-like dog, outside unattended, assuming the noisy fox would take the hint and leave.

That wasn't quite what happened.

Other local people stopped their chores to stare in bewilderment as a fox and a small orange dog trotted together down the street, apparently quite at ease with each other. It happened again and again – night after night, they were a fixture on the village scene. The dog's owner concluded that the fox's barking was actually a call for her 'friend' to join her.

When it comes to bonding between unlikely animals, this equalled anything dreamed up in Disney's heyday.

Terriers were bred to kill wildlife, not to amble down the street with it.

Why would a fox travel with a dog? They rarely hunt in the company even of their own kind, let alone a potentially lethal enemy. Perhaps having a dog as a companion provides a competitive advantage against other foxes.

Or perhaps not all rare behaviour requires an explanation.

Perhaps there is still that corner of a fox's mind that remains beyond our ability to understand.

The Fox Watcher's Toolkit

If a science teacher were to request a study species to demonstrate the natural world, his plea might be answered with a fox. In one slender orange frame, they provide an intriguing wild package of animal behaviour, ecological interactions, and human–wildlife relationships which is accessible, real and dynamic – which is more than can be said, at least in south-east England, for tigers, pandas or elephants. Foxes are a signpost that can guide our increasingly gadget-centric culture back to respect and awareness of the natural universe, and, perhaps, a better understanding of humanity's place within it.

This chapter lists a few activities that take advantage of the fox's presence in our midst.

Photography

We might live in the age of the selfie but capturing pictures of other living creatures has never been more popular. It is also fair to acknowledge that it can be very addictive. What begins with a phone camera otherwise used for babies and Christmas parties can escalate into an endless variety of digital single lens reflex (DSLR) lenses, tripods and Adobe editing software. But at either end of the spectrum, it is a very rewarding hobby that makes even the most seasoned wildlife-watcher more observant and appreciative of their surroundings.

One caveat: the animal's welfare must come first. Most old-timers in the photography game can tell bizarre stories of watching phone-wielding tourists walking up to wild grizzly bears or getting chased by angry stags, but apart from human safety issues, there is also a risk of frightening the subject and making it abandon its food or offspring. Parks Canada's maxim is useful everywhere: if you make the animal change its behaviour, you are too close. When stalking wildlife with my camera, I do my best to keep to that philosophy. The perfect photo is one obtained without the animal even knowing that I am there.

What kind of camera is best for foxes? The first thing to consider is that you will need a considerable zoom. I use a 200–500 mm lens on my DSLR; on a fixed lens camera, at least a 10x zoom is useful. This means that blurred pictures through camera shake can be a real problem, worsened by the fact that foxes often appear in low light, necessitating a slow shutter speed. Of course, there will always be one fox who strolls across the lawn at midday, but it is best to assume that most pictures won't be taken in ideal conditions.

There is no upper limit to the expense of wildlife photography; it can consume however much money you decide to invest in it. Fortunately, there is no end to the creative potential either. I haven't yet managed a picture of a vixen and four cubs trotting in silhouette under a golden spring sunrise, but every opportunity for a photograph gives a little more insight into their world.

Trail camera (camera trap) photography

Like regular photography, this activity can be as expensive as you desire, although prices have come down somewhat in recent years. The idea is that you tie a specialised motion-sensitive camera to a tree or fencepost, and leave it there long enough for something wild to trigger it. There was a time when buying a trail camera meant importing it from the USA and braving custom duties, but these days they are easily available through Amazon, eBay, and other retailers large and small. Reconyx, Bushnell and Acorn are among the most popular models, although I have to acknowledge Browning as being possibly the best balance

between price, reliability and quality. If you really want a trail camera for a bargain, Amazon has many cheap brands, but in my experience they tend to stop working after a few months.

I have used trail cameras in some of the world's most remote wildernesses for serious scientific research, but even when not capturing prides of lions in the Serengeti, they are undeniably great fun. Depending on your luck, a trail camera at the end of the garden might catch foxes making a 2 a.m. visit, an unexpected badger, or your neighbour's tabby. Of course, the wilder the setting, the greater the chance of photographing a shy species, although be aware of landowner permission requirements. There is also an inherent security risk to leaving a camera alone in a forest for a week. Specialised steel security cases for trail cameras are now available, which can be secured with a Python cable or high quality chain. Bicycle locks won't last a minute in the hands of a thief – I learned this the hard way. Regardless of the cable type, be sure to buy a high-grade padlock as well. However, undoubtedly the best safety measure for a trail camera is to conceal it as much as possible from public view.

If foxes regularly visit your garden, a trail camera will probably give new insights into their relationships and activity patterns, and might even reveal the presence of transients coming at lonely hours to avoid the dominant pair. You may catch some interesting behaviour too. Fighting, mating and playing with stolen shoes are much more likely to be observed by a trail camera on constant duty than an intermittent human watcher.

Most trail cameras will give you a choice between still photographs and movies. The majority use infrared for producing night shots; there are still a few models that use a white flash, which produces an attractive colour image but risks higher disturbance to the animal. And, of course, it is more likely to be spotted by people; the white beam can be seen from afar.

I typically set trail cameras to shoot movies for thirty seconds. This is a compromise between getting a nice clip and not wasting hours of card space on footage of grass being shaken by the wind. And you will have misfires by the dozen, when the animal is just out of range, or the camera has been triggered by weather. If the camera is in a public place, you also run the risk of accidentally filming people, some of whom will realise this fact and react in remarkably odd ways.

A good trail camera site, then, is one with minimal use by humans and dogs. For foxes, a camera should be at eye-level – about 45 cm off the ground – and looking down a wildlife path rather than across it. Areas exposed to wind-shaken grass or branches are best avoided.

Tracking

Humans have possessed cameras for only a blink on the timeline of our history, but tracking is as old as our species. Motivations may have changed since the Stone Age, but the quiet Sherlockian thrill of piecing together subtle evidence in the landscape is still one with great appeal.

Almost any surface can reveal clues to an experienced observer, but snow is of course the easiest medium. Fox paws are narrower than a cat's and have a smaller heel pad than a dog's, but the easiest way to distinguish them is that the trail will be neat and straight. If it meanders, or if human tracks are parallel to it, it is probably the work of a small dog. In a good quality print, you will notice that the

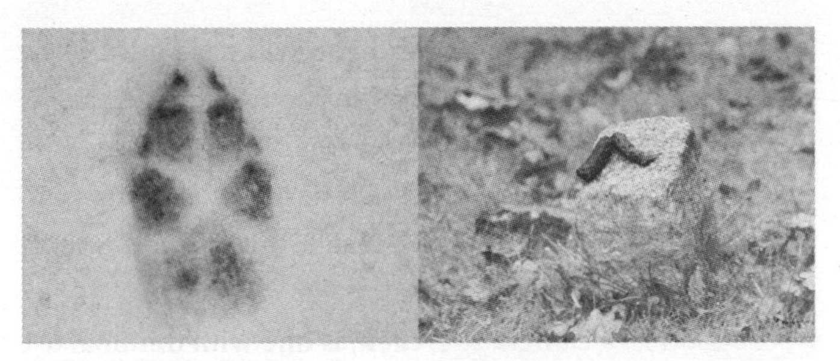

Fox footprint and scat.

two front toes are set slightly ahead of the two back toes, to the extent that a horizontal line can be drawn between them without crossing any pads. In dogs, the track is rounder, and the front toes are only just offset, so no line is possible.

Foxes are usually very loyal to their paths, and over time a track of slightly lower grass will indicate their movements. If this track is very worn – down to the earth – it may well indicate use by badgers too. Badger pawprints, for the record, have five toes in a straight line along the top of a huge kidney-shaped heel pad; in poor conditions, the fifth toe may not register, but the track is always much wider than a fox's neat print. With either species, try following the trail until it ducks under brambles or a barbed-wire fence. Can you see any snags of fur? Fox hair will not necessarily be orange, given how much white and grey is in their pelt, but it will be soft. Badger hair is angular and feels ridged when rolled between your fingers.

Other signs will be identified through experience. Fox scat is smaller, darker and more twisted than that of most dogs, and will often contain berries or fur, or be grey with crunched-up bones. It may be placed on anthills or rocks as a scent message to other foxes, or on discarded toys and emptied caches. It smells musty-sweet, and their urine is perceptible to an alert human nose long after it has been deposited.

A heap of woodpigeon feathers in a park may be the work of a fox, sparrowhawk, or unfortunately a cat. A fox plucks a pigeon clumsily, biting through the primary feathers and leaving the quill cut neatly as if with scissors, while a hawk hooks them out with its curved bill, leaving distinct holes.

Taking casts of tracks

This is surprisingly easy and is a good activity to share with children. You will of course need a clear fox track in reasonably firm mud or sand. If foxes regularly visit your garden, so much the easier; otherwise, have a look at the tracking tips in the previous section.

Apart from a footprint, you will also need plaster of Paris, water, and cardboard or thick plastic cut into four strips about 10 cm long and 4 cm wide. Arrange the strips to form a rectangle around the track, each piece upright on its long side, preferably pushed into the mud for stability.

Mix up the plaster of Paris and the water, stirring until it is smooth. Then, pour it over the track until the liquid fills the rectangle.

It should take about fifteen minutes to set, dependent on weather and the mixture's consistency. Once it is firm, remove it from the earth, digging well under the cast to avoid breaking it. The best trick is to use a shovel or garden fork. You will find yourself lifting a kind of cake out of the ground, with a dense layer of mud and the cast on top. If the cast is fully set, the mud can be washed off without damaging the print, although some brushing may be needed to get rid of the final specks. Hopefully, if it has not cracked during the cleaning exercise, you'll have a nice wildlife memento with which to decorate your house or schoolroom.

Keeping a wildlife diary

If the paper version seems too Victorian, a blog or Twitter account will chronicle records of your sightings just as effectively, and can allow you to share them with an audience of other wildlife fans. Normal safety guidelines for sharing personal information online obviously apply – in fact, they are heightened, because giving sighting locations on social media can be very dangerous to the animal.

That being said, your own private records should contain such details as date, time, weather and number of individuals observed. Even if your interest is primarily in foxes, please note any rare species that you happen across. Your local wildlife records centre will welcome the information.

Casual scientific projects

At its heart, science is curiosity. Merely noting the presence or absence of a fox on your front lawn is in essence a scientific action. But those sightings become more enjoyable as patterns are determined. Even if you do not wish to sign up for a PhD and write a 250-page thesis, your fox-watching can still involve some research.

Most science, like most stories, looks at the contrasts and relationships between two or more variables or characters. So what to compare? Fortunately, there's one crucial variable that requires no special equipment to measure: *time*. Think about the year from the fox's perspective; every season brings its own challenges and opportunities.

Over the course of a year, patterns will start to emerge. Why not start with recording all fox sounds that you hear? Learn the different calls – screams, chirps, courtship barks – and see how they change over time. Alternatively, if you consistently travel the same route at the same time of day – for example, on a train commute, or walking your dog – you may also notice trends.

If you have a group of foxes visiting your garden, try to recognise at least some individuals. Look for muzzle scars, chipped ears, distinctive brush tips, and general face shape. Distinguishing foxes is arguably easier than telling apart two tabby cats, and most pet owners don't struggle with that. Recognition of individuals makes fox-watching far more rewarding. Not only will you start to learn the unique personality of each fox, but you will also be able to make sense of their interactions with each other. Who is dominant over whom? Are some foxes greeted and others chased away? Is last year's breeding vixen lactating again this spring?

With persistence and some luck, you can piece together their stories.

Science projects for students

For students seeking a research project for their A-levels or university dissertation, foxes offer almost unlimited potential. Here are just a few ideas:

- Interactions between foxes and cats in suburban gardens: are most encounters neutral, cat-dominated or fox-dominated?
- How does a scientific fox density estimate in a town park compare to the anecdotal estimates of 100 local people?

- Diet analysis of rural and urban fox scats: does anthropogenic food abundance differ?
- To what degree do mange records from a town for one year predict outbreaks in following years? (Wildlife rescue centres may be able to provide data.)
- What is the science content of 100 recent news articles about foxes, and how does this compare between tabloids, broadsheets, television and internet news?

More complex and expensive projects:

- By what mechanism do badgers reduce fox abundance?
- What factors predict an individual fox's susceptibility to mange: location, habitat, diet, age, exposure to rodenticides, etc.?
- Is fox predation on roe and muntjac deer additive mortality in the UK? Does this vary between arable farms and ancient woodland?
- Can the impact on fox density of removing deliberate feeding from a London borough be modelled?
- How does fox dispersal of plant seeds influence the species richness and distribution of woodland flora?

- What are the most effective methods of reducing road mortality in foxes and other canids, and how do these vary between road type?

Resources

My corner of the web:

www.adelebrandblog.wordpress.com/

www.facebook.com/awalkwithwildlife

www.youtube.com/sittingfox

Useful sites:

Wildlife Online www.wildlifeonline.me.uk

Help Wildlife (contacts for emergencies) www.helpwildlife.co.uk

Canid Specialist Group (International Union for the Conservation of Nature) www.canids.org

The Fox Website (University of Bristol) www.thefoxwebsite.net

ACKNOWLEDGEMENTS

WITH PARTICULAR THANKS TO Marc Baldwin for proofreading, to my parents for enjoying their garden being turned into a fox observation and rescue depot, and to Ann Lardeur for sharing her entertaining fox stories. Thanks also to the many British and international visitors who drop by my website and YouTube channel. I apologise for deafening some of you with the White Socks Vixen's unmistakable voice.

Finally, I would like to thank the foxes themselves for crossing my path over the last 25 years – from the windswept prairies of Canada to the sweet meadows of Romania, and from jaguar-prowled jungles in Mexico to the scorching salt deserts of India. They make the world a more mysterious and interesting place.

BIBLIOGRAPHY

Chapter 1

Stanmore, I. Undated. Disappearance of wolves in the British Isles. Wolf Song of Alaska website http://www.wolfsong alaska.org/chorus/node/230

BBC. 30 March 2015: Election 2015: Fox makes a bid for Downing Street

http://www.bbc.co.uk/news/election-2015-32112635

Huffington Post. 5 March 2002: QPR's squirrel and the greatest animal pitch invaders. https://www.huffington post.co.uk/2012/03/05/qprs-squirrel-and-the-best-animal-pitch-invaders_n_1320390.html?guccounter=1

Conniff, R. 10 November 2015. Learning to live with leopards. National Geographic http://ngm.nationalgeographic.com/2015/12/leopards-moving-to-cities-text

Urban Hyaena Research Project. 2012. University of Bristol. http://www.wilddogconservationmalawi.org/urban hyaena.html

Huffington Post. 15 January 2016. Cougar In North Vancouver Backyard Came Out Of Nowhere. www.huffingtonpost.ca/2016/01/15/cougar-north-vancouver-video_n_8995360.html

Chapter 2

Voorspoels, S; Covaci, A; Lepom, P; Escutenaire, S; Schepens, P. 2006. Remarkable Findings Concerning PBDEs in the Terrestrial Top-Predator Red Fox (*Vulpes vulpes*). *Environmental Science and Technology* **40**(9): 2937–43

Lowe, V P; Horrill, A D. 1991. Caesium concentration factors in wild herbivores and the fox (*Vulpes vulpes L*). *Environmental Pollution* **70**(2): 93–107

Hilbers, D; ten Cate, B. 2013. *North-east Poland*. Crossbill Guides, Arnhem, The Netherlands

Jędrzejewski, W; Jędrzejewski, B. 1992. Foraging and diet of the red fox *Vulpes vulpes* in relation to variable food resources in Białowieża National Park, Poland. *Ecography* **15**(2): 212–20

Wilmers, C C; Stahler, D R; Crabtree, R L; Smith, D W; Getz, W M. 2003. Resource dispersion and consumer dominance: scavenging at wolf- and hunter-killed carcasses in Greater Yellowstone, USA. *Ecology Letters* **6**: 993–1003

Yalden, D. 1999. *The History of British Mammals*. Academic Press, London

Wang, X; Telford, R H. 2008. *Dogs: Their fossil relatives and evolutionary history*. Columbia University Press, Chichester, West Sussex

de Bonis, L; Peigné, S; Likius, A; Mackaye, H T; Vignaud, P; Brunet, M. 2007. The oldest African fox (*Vulpes riffautae n. sp., Canidae, Carnivora*) recovered in late Miocene deposits of the Djurab desert, Chad. *Naturwissenschaften* **94**(7): 575–80

Janossy, D. 1986. Pleistocene vertebrate faunas of Hungary. *Developments in Palaeontology and Stratigraphy* **8**. Elsevier,

Amsterdam 1-208 Quoted on: Fossilworks.org http://fossilworks.org/bridge.pl?action=collectionSearch&collection_no=35369&is_real_user=1 Website visited 25 April 2016

Sommer, R; Benecke, N. 2005. Late-Pleistocene and early Holocene history of the canid fauna of Europe (Canidae). *Mammalian Biology – Zeitschrift für Säugetierkunde* **70**(4): 227–41

Worldwide Fund for Nature. 2008. Climate change likely culprit as Arctic fox faces extinction. http://wwf.panda.org/wwf_news/?147581/Climate-change-likely-culprit-as-arctic-fox-faces-extinction

Maher, L A; Stock, J T; Finney, S; Heywood, J J N; Miracle, P T; Banning, E B. 2011. A Unique Human-Fox Burial from a Pre-Natufian Cemetery in the Levant (Jordan). *PLoS ONE* **6**(1): e15815. doi:10.1371/journal.pone.0015815

Science daily. 2 February 2011. Anthropologists discover earliest cemetery in Middle East. https://www.sciencedaily.com/releases/2011/02/110202132609.htm

Horwitz, L K; Goring-Morris, N. 2004. Animals and ritual during the Levantine PPNB : a case study from the site of Kfar Hahoresh, Israel. *Anthropozoologica* **39**(1): 165–78

Curry, A. 2008. Gobekli Tepe: The World's First Temple? Smithsonian Magazine Website http://www.smithsonianmag.com/history/gobekli-tepe-the-worlds-first-temple-83613665/?no-ist

Jicarilla Apache. Undated. The Origin of Fire. http://sfrc.ufl.edu/plt/activities_files/Origin_of_Fire.pdf

Nozaki, K. 1961. *Kitsune: Japan's Fox of Mystery, Romance, and Humor*. Hokuseido Press. Printed in Japan

Spaeth, B S. 1996. *The Roman Goddess Ceres*. University of Texas Press, pp 36–7.

Varty, K. 2000. *Reynard the Fox: Social engagement and cultural metamorphoses in the beast epic from the Middle Ages to present*. Berghahn Books, Oxford, p. 163

Freeborn, D. 2006. *From Old English to Standard English*. Palgrave Macmillan, Hampshire, p. 137

Statham, M J; Murdoch, J; Janecka, J; Aubry, K B; Edwards, C J; Soulsbury, C D; Berry, O; Wang, Z; Harrison, D; Pearch, M; Tomsett, L; Chupasko, J; Sacks, B N. 2014. Range-wide multilocus phylogeography of the red fox reveals ancient continental divergence, minimal genomic exchange and distinct demographic histories. *Mol Ecol* **23**: 4813–30

Chapter 3

National Field Atlas. Mammal Society. https://data.nbn.org.uk/Taxa/NHMSYS0000080188 Website visited 25 April 2016

Harris, S; Yalden, D W. 2008. *Mammals of the British Isles Handbook*. Published by the Mammal Society c/o Society for Experimental Biology, Southampton

Jones, D M; Theberge, J B. 1982. Summer home range and habitat utilisation of the red fox (*Vulpes vulpes*) in a tundra habitat. *Northwest British Columbia Canadian Journal of Zoology* 60(5): 807–12

Sidorovich, V E; Sidorovich, A A; Izotova, I. 2006. Variations in the diet and population density of the red fox *Vulpes Vulpes* in the mixed woodlands of northern Belarus. *Mammalian Biology* **72**(2): 74–89

Meia, J S; Weber, J M. 1995. Home ranges and movements of red foxes in central Europe: stability despite environmental changes. *Canadian Journal of Zoology* 73(10): 1960–6

Adkins, C A; Stott, P. 1998. Home ranges, movements and habitat associations of red foxes *Vulpes Vulpes* in suburban Toronto, Ontario, Canada. *Journal of Zoology* **244**(3): 335–46

Goszczyński, J. (2002). Home ranges in red fox: territoriality diminishes with increasing area. *Acta Theriologica* **47**: 103–14

Taylor, I. 1994. *Barn Owls: Predator-prey relationships and conservation*. Cambridge University Press, Cambridge, UK, p. 73

Vuroisalo, T; Talvitie, K; Kauhala, K; Blauer, A; Lahtinen, R. 2014. Urban red foxes (*Vulpes Vulpes*) in Finland: A historical perspective. *Landscape and Urban Planning* **124**: 109–17

Nozaki, K. 1961. *Kitsune Japan's Fox of Mystery, Romance, and Humor*. Hokuseido Press. Printed in Japan

Baumann, C; Starkovich, B M; Bocherens, H; Conard, N J. 2016. Food strategies of red and arctic foxes during the pre-Last Glacial Maximum (Ach Valley, Swabian Jura, Germany). Poster submitted to StEvE Meeting (November 2016)

Trewhella, W J; Harris, S. 1990. The effect of railway lines on urban fox (*Vulpes vulpes*) numbers and dispersal movements. *Journal of Zoology* 221(2): 321–6

Harris, S; Rayner, J M V. 1986. A Discriminant Analysis of the Current Distribution of Urban Foxes (*Vulpes vulpes*) in Britain. *Journal of Animal Ecology* **55**(2): 605–11

Kolb, H. 1984. Factors Affecting the Movements of Dog Foxes in Edinburgh. *Journal of Applied Ecology* **21**(1): 161–73

Galov, A; Sindicic, M; Andreanszky, T; Curkovic, S; Dezdek, D; Slavica, A; Hartl, GB; Krueger, B. 2014. High genetic diversity and low population structure in red foxes (*Vulpes vulpes*) from Croatia. *Mammalian Biology* **79**: 77–80

Chapter 4

Červený, J; Begall, S: Koubek, P; Nováková, P; Burda, H. Directional preference may enhance hunting accuracy in foraging foxes. *Biology Letters*. Published online before print 12 January 2011

Wikipedia. Domestic cat. en.wikipedia.org/wiki/Cat. Website visited 7 September 2017

Henry, J D. 1996. *Red Fox: The Catlike Canine*. Smithsonian Institution Press, Washington DC

Makemper, E P; Topinka, V; Burda, H. 2015. A behavioral audiogram of the red fox (*Vulpes vulpes*). *Hearing Research* **320**: 30–7

Banks, M S; Sprague, W W; Schmoll, J; Parnell, J A Q; Love, G D. 2015. Why do animal eyes have pupils of different shapes? *Science Advances* **1**(7): e1500391 DOI: 10.1126/sciadv.1500391

Harris, S; Yalden, D W. 2008. *Mammals of the British Isles Handbook*. Published by the Mammal Society c/o Society for Experimental Biology, Southampton

Macgillivray, W. 1838. *A History of British Quadrupeds*. W H Lizars, Edinburgh, p. 183

Chapter 5

Theberge, J; Theberge, M. 2013. *Wolf Country: Eleven Years Tracking the Algonquin Wolves*. McClelland & Stewart, Toronto

Cederlund, G; Linstrom, E. 1983. Effects of severe winters and fox predation on roe deer mortality. *Acta Theriologica* **28**(7): 129–45

Radinsky, L. 1973. Evolution of the Canid Brain. *Brain, Behavior and Evolution* 7(3): 186–202

Obidzinski, A; Kieltyk, P. 2006. Changes in ground vegetation around badger setts and fox dens in the Białowieża Forest, Poland. *Polish Botanical Studies* **22**: 407–16

Harris, S. 1979. Age-related fertility and productivity in Red foxes, *Vulpes vulpes*, in suburban London. *Journal of Ecology* **187**(2): 195–9

Jenness, R. 1979. The composition of human milk. *Semin Perinatol* 3(3): 225–39

Harris, S; Yalden, D W. 2008. *Mammals of the British Isles Handbook*. Published by the Mammal Society c/o Society for Experimental Biology, Southampton

Henry, J D. 1996. *Red Fox: The Catlike Canine*. Smithsonian Institution Press, Washington DC

Dagg, A. Infanticide by Male Lions Hypothesis: A Fallacy Influencing Research into Human Behavior. *American Anthropologist* **100**(4): 940–50

Baker, P J; Robertson, C P J; Funk, S M; Harris, S. 1998. Potential fitness benefits of group living in the red fox, *Vulpes vulpes*. *Animal Behaviour* **56**(6): 1411–24

Von Schantz, T. 1981. Female cooperation, male competition, and dispersal in the red fox *Vulpes vulpes*. *Oikos* **37**: 63–8

Hartley, F G L; Follett, B K; Harris, S; Hirst, D; McNeilly, A S. 1994. The endocrinology of gestation failure in foxes (*Vulpes vulpes*). *Journal of Reproduction and Fertility* **100**: 341–6

Page, R J C. 1981. Dispersal and population density of the fox (*Vulpes vulpes*) in an area of London. *Journal of Zoology* **194**(4): 485–91

BBC. Urban fox's record-breaking country walk. http://www.bbc.co.uk/nature/25759153 Website visited 28 September 2017

Gosselink, T; Piccolo, K, Van Deelen, T; Warner, R. 2010. Natal Dispersal and Philopatry of Red Foxes in Urban and Agricultural Areas of Illinois. *Journal of Wildlife Management* **74**(6): 1204–17

Chapter 6

Juan, T; Sagrario, A; Jesus, H; Cristina, C M. 2006. Red fox (*Vulpes Vulpes L.*) favour seed dispersal, germination and seedling survival of Mediterranean Hackberry (*Celtis australis L.*). *Acta Oecologica* **30**(1): 39–45

Obidzinski, A; Kieltyk, P. 2006. Changes in ground vegetation around badger setts and fox dens in the Białowieża Forest, Poland. *Polish Botanical Studies* **22**: 407–16

Satchell, J E. 1983. *Earthworm Ecology: From Darwin to Vermiculture.* Chapman and Hall, New York, p. 404

Mail Online. 16 October 2010. Bloodbath at London Zoo as urban foxes slaughter 11 penguins and one flamingo. http://www.dailymail.co.uk/news/article-1321099/

Bloodbath-London-Zoo-caused-urban-foxes-report-reveals.html Website visited 2 May 2016

Henry, J D. 1996. *Red Fox: The Catlike Canine*. Smithsonian Institution Press, Washington DC, p. 63

European Commission. 2009. *European Union Management Plan 2009–2011: Lapwing Vanellus vanellus*. Luxembourg: Office for Official Publications of the European Communities, p. 29

Seymour, A S; Harris, S; Ralston, C; White, P C L. 2003. Factors influencing the nesting success of Lapwings *Vanellus vanellus* and behaviour of Red Fox *Vulpes vulpes* in Lapwing nesting sites. *Bird Study* **50**(3): 39–46

Defra. Undated. Project Code DB1327: Management of wet grassland habitat to reduce the impact of predation on breeding waders. http://randd.defra.gov.uk/Document.aspx?Document=BD1327_10170_FRP.doc Website visited 2 May 2016

Macdonald, D W. 1994. Behavior of Red Foxes, *Vulpes vulpes*, Caching Eggs of Loggerhead Turtles, *Caretta caretta*. *Journal of Mammology* 75(4): 985–8

Henry, J D. 1996. *Red Fox: The Catlike Canine*. Smithsonian Institution Press, Washington DC, p. 88

Harris, S; Yalden, D W. 2008. *Mammals of the British Isles Handbook*. Published by the Mammal Society c/o Society for Experimental Biology, Southampton, p. 105

Barn Owl Trust. 2012. *Barn Owl Conservation Handbook: A comprehensive guide for ecologists, surveyors, land managers and ornithologists*. Pelagic Publishing, Exeter. Section 2.2.1. Pages not numbered

Villar, N; Lambin, X; Evans, D; Pakeman, R; Redpath, S. 2013. Experimental evidence that livestock grazing intensity affects the activity of a generalist predator. *Acta Oecologica* **49**: 12–16

Bell, D. Senior Lecturer at University of East Anglia. Email 24 February 2016

Bradford, W. Wildlife conflict specialist, Parks Canada, Jasper National Park. Email 10 December 2004

Jarnemo, A; Liberg, O; Lockowandt, S; Olsson, A; Wahlström, K. 2004. Predation by red fox on European roe deer fawns in relation to age, sex, and birth date. *Canadian Journal of Zoology* **82**(3): 416–22

O'Donoghue, A. 1991. Growth, Reproduction and Survival in a Feral Population of Japanese Sika Deer (Cervus nippon) Unpublished PhD thesis, University College, Dublin, Ireland. In: Putman, R J. 2008. A review of available data on natural mortality of red and roe deer populations. Commissioned by the Deer Commission for Scotland

Cederlund, G; Linstrom, E. 1983. Effects of severe winters and fox predation on roe deer mortality. *Acta Theriologica* **28**(7): 129–45

Jarnemo, A; Liberg, O. 2005. Red fox removal and roe deer fawn survival – a 14-year study. *Journal of Wildlife Management* **69**(3): 1090–8

Jędrzejewski, W; Jędrzejewska, B. 1992. Foraging and diet of the red fox *Vulpes vulpes* in relation to variable food resources in Białowieża National Park, Poland. *Ecography* **15**(2): 212–20

Young, A; Márquez-Grant, N; Stillman, R; Smith, M J;

Korstjens, A. 2013. An investigation of red fox (*Vulpes vulpes*) and Eurasian badger (*Meles meles*) scavenging, scattering and removal of deer remains: forensic implications and applications. *Journal of Forensic Science* **60** Suppl 1:S39-55

Henry, J D. 1996. *Red Fox: The Catlike Canine*. Smithsonian Institution Press, Washington DC, pp 96, 111

Trewby, I D; Wilson, G J; Delahay, R J; Walker, N; Young, R; Davison, J; Cheeseman, C; Robertson, P A; Gorman, M L; McDonald, R A. 2008. Experimental evidence of competitive release in sympatric carnivores. *Biology Letters* 4(2): 170–2

Macdonald, D W; Buesching, C D; Stopka, P; Henderson, J; Ellwood, S A; Baker, S E. 2004. Encounters between two sympatric carnivores: red foxes (*Vulpes vulpes*) and European badgers (*Meles meles*). *Journal of Zoology* **263**(4): 385–92

Newsome, T M; Ripple, W J. 2015. A continental scale trophic cascade from wolves through coyotes to foxes. *Journal of Animal Ecology* **84**(1): 49–59

Carbone, C; Mace, G M; Roberts, S C; Macdonald, D W. 1999. Energetic constraints on the diet of terrestrial carnivores. *Nature* **402**: 286–8

Murie, A. 1944 (paperback edition 1985, 2001). *The Wolves of Mount McKinley*, University of Washington Press, Seattle, pp 219–20

Baker, P J; Harris, S. 2006. Does culling reduce fox (*Vulpes vulpes*) density in commercial forests in Wales? *European Journal of Wildlife Research* **52**: 99–108

Theuerkauf, J; Rouys, S; Jędrzejewski, W. 2003. Selection of den, rendezvous, and resting sites by wolves in the Białowieża Forest, Poland. *Canadian Journal of Zoology* **81**: 163–7

Lynx UK Trust. 11 March 2015. Untitled online article https://www.facebook.com/lynxuktrust/posts/798292100220484:0. Website visited 2 May 2016

Chapter 7

Jennings, D J; Carlin, C M; Gammell, M P. 2009. A winner effect supports third-party intervention behaviour during fallow deer, *Dama dama*, fights. *Animal Behaviour* 77(2): 343–8

Shabadash, S A; Zelikina, T I. 2004. The Tail Gland of Canids. *Biology Bulletin* **31**: 367

Daily Nebraskan. Campus evergreens sprayed with fox urine to prevent theft http://www.dailynebraskan.com/news/campus-evergreens-sprayed-with-fox-urine-to-prevent-theft/article_8640fa46-6d53-11e5-b6be-1706586 e9c62.html, 2 May 2016

Scheinin, S; Yom-Tov, Y; Motro, U; Geffen, E. 2006. Behavioural responses of red foxes to an increase in the presence of golden jackals: a field experiment. *Animal Behaviour* **71**: 577–84

Dickman, C R; Doncaster, C P. 1984. Responses of small mammals to Red fox (*Vulpes vulpes*) odour. *Journal of Zoology* **204**(4): 521–31

Henry, J D. 1996. *Red Fox: The Catlike Canine*. Smithsonian Institution Press, Washington DC

Chapter 8

University of Bristol. Sarcoptic mange in red foxes: the role of fox behaviour. http://www.bio.bris.ac.uk/research/mammal/fox_mange.html Website visited 28 September 2017

Daily Telegraph. 13 June 2013. Let's have foxhunting in London, says Boris. https://www.telegraph.co.uk/news/politics/conservative/10117433/Lets-have-fox-hunting-in-London-says-Boris.html Website visited 30 March 2019

The Fox. Foxhunting: trends in fox numbers. http://www.thefoxwebsite.net/foxhunting/hunttrends Website visited 28 September 2017

DEFRA. Red fox *Vulpes vulpes*. http://www.mammal.org.uk/sites/default/files/DEFRA%20red%20fox%20research_1.pdf Website visited 28 September 2017

Chapter 9

Get Surrey. Man facing court over gin trap fox deaths in Guildford. http://www.getsurrey.co.uk/news/surrey-news/man-facing-court-over-gin-8437534 Website visited 29 September 2017

Crown Prosecution Service. Guide to Wildlife Offences. http://www.cps.gov.uk/legal/v_to_z/wildlife_offences/ Website visited 29 September 2017

Centres for Disease Control and Prevention. Alveolar Echinococcosis FAQ. https://www.cdc.gov/parasites/echinococcosis/gen_info/ae-faqs.html Website visited 29 September 2017

Learmount, J; Zimmer, A; Conyers, C; Boughtflower V D; Morgan, C P; Smith, G C. 2012. A diagnostic study of *Echinococcus multilocularis* in red foxes (*Vulpes vulpes*) from Great Britain. *Veterinary Parasitology* **190**(3–4): 447–53

Craig, P S; Woods, M L; Boufana, B; O'Loughlin, B; Gimpel, J; Lett, W S; McManus, D P. 2012. Cystic echinococcosis in a fox-hound hunt worker, UK. *Pathogens and Global Health* **106**(6): 373–5

Centres for Disease Control and Prevention. Rabies. https://www.cdc.gov/rabies/ Website visited 29 September 2017

World Health Organisation. Rabies Factsheet. http://www.who.int/mediacentre/factsheets/fs099/en/ Website visited 29 September 2017

BBC News. Q&A Rabies. http://www.bbc.co.uk/news/health-18188682 Website visited 29 September 2017

Zimen, E. 1980. The Red Fox: Symposium on Behaviour and Ecology. *Biogeographica* **18**. Springer Science Business Media, The Hague, p. 238

Sadhowska-Todys, M; Kucharczyk, B. 2014. Rabies in Poland in 2012. *Przegląd epidemiologiczny* **68**(3):465–8, 567–9

Andriantsoanirina, V; Fang F; Ariey, F; Izri A; Foulet, F; Botterel, F; Bernigaud, C; Chosidow, O; Huang, W; Guillot, J; Durand, R. 2016. Are humans the initial source of canine mange? *Parasites & Vectors* **9**: 177

Currier R W; Walton S F; Currie B J. 2011. Scabies in animals and humans: history, evolutionary perspectives, and modern clinical management. *Annals of the New York Academy of Sciences* **1230**: E50–E60

Pence, D B; Ueckermann, E. 2002. Sarcoptic mange in wildlife. *Revue scientifique et technique* **21**(2): 385–98

NHS Choices. Scabies – causes. http://www.nhs.uk/Conditions/Scabies/Pages/Causes.aspx Website visited 29 September 2017

Soulsbury, C D; Iossa, G; Baker, P J; Cole, N C; Funk, S M; Harris, S. 2007. The impact of sarcoptic mange *Sarcoptes scabiei* on the British fox *Vulpes vulpes* population. *Mammal Review* **37**(4): 278–96

Nimmervoll, H; Hoby, S; Robert, N; Lommano, E; Welle, M; Ryser-Degiorgis, P R. 2013. Pathology of sarcoptic mange in red foxes (*Vulpes vulpes*): macroscopic and histologic characterization of three disease stages. *Journal of Wildlife Diseases* **49**(1): 91–102

Fox Wood Wildlife Rescue. Treating Sarcoptic Mange in Red Foxes. http://www.foxwoodwildliferescue.org/2017/01/05/treating-sarcoptic-mange-in-red-foxes/ Website visited 29 September 2017

Tolhurst, B; Grogan, A; Hughes, H; Scott, D. 2016. Effects of temporary captivity on ranging behaviour in urban red foxes (*Vulpes vulpes*). *Applied Animal Behaviour Science* **181**: 182–90

National Fox Welfare Society. http://www.nfws.org.uk/mange/ Website visited 29 September 2017

Guardian. A Kind of Magic? 2007. www.theguardian.com/science/2007/nov/16/sciencenews.g2 Website visited 29 September 2017

Wired. 2011. Skeptic offers $1 million for proof that homeopathy works. http://www.wired.co.uk/article/homeopathy-challenge Website visited 29 September 2017

Mulder, J. 2004. Longevity records in the red fox. *Lutra* **47**(1): 51–2

Chapter 10

Guinness Book of Records. First documented animal execution. http://www.guinnessworldrecords.com/world-records/first-documented-animal-execution Website visited 29 September 2017

Outside Online. 5 April 2016. The truth about wolf surplus killing: survival, not sport. https://www.outsideonline.com/2066881/truth-about-wolf-surplus-killing-survival-not-sport Website visited 29 September 2017

Kruuk, H. 1972. Surplus killing by carnivores. *Journal of Zoology* **166**(2): 233–44

Rabbit Rehome. http://www.rabbitrehome.org.uk/care/rabbit_foxes.asp Website visited 11 September 2017

Metropolitan Police. 20 September 2018. Scavenging by wildlife established as likely cause of reported cat mutilations. http://news.met.police.uk/news/scavenging-by-wildlife-established-as-likely-cause-of-reported-cat-mutilations-323426 Website visited 17 October 2018

Royal Horticultural Society. Fox. https://www.rhs.org.uk/advice/profile?PID=511 Website visited 29 September 2017

Chapter 11

BBC News. Mother's 'nightmare' after twins 'mauled' by fox. http://www.bbc.co.uk/news/10251349 Website visited 29 September 2017

Mail Online. 8 June 2010. Fox attack on my girls was like horror film: Mother relives nightmare moment she found her twins mauled in their cots. http://www.dailymail.co.uk/news/article-1284505/Baby-twins-Isabella-Lola-Koupparis-seriously-injured-fox-attack.html Website visited 29 September 2017

White, P C L; Harris, S. 1994. Encounters between red foxes (*Vulpes vulpes*): implications for territory maintenance, social cohesion and dispersal. *Journal of Animal Ecology* **63**(2): 315–27

Chapter 12

Black Foxes. http://www.blackfoxes.co.uk/ Website visited 29 September 2017

Bored Panda. Fox Village In Japan Is Probably The Cutest Place On Earth. https://www.boredpanda.com/zao-fox-village-japan/ Website visited 29 September 2017

Science. 2015. How dogs stole our hearts. http://www.sciencemag.org/news/2015/04/how-dogs-stole-our-hearts Website visited 29 September 2017

Galibert, F; Quignon, P; Hitte, C; Andre, C. 2011. Toward understanding dog evolutionary and domestication history. *Comptes Rendus Biologies* **334**(3):190–6

National Geographic. 2013. Opinion: We Didn't Domesticate Dogs. They Domesticated Us. http://news.nationalgeographic.com/news/2013/03/130302-dog-domestic-evolution-science-wolf-wolves-human/ Website visited 29 September 2017

BBC Earth. A Soviet scientist created the only tame foxes in the world. http://www.bbc.co.uk/earth/story/20160912-a-soviet-scientist-created-the-only-tame-foxes-in-the-world Website visited 29 September 2017

Globe and Mail. 10 January 2010. Exotic animal owner killed by 650-pound tiger https://beta.theglobeandmail.com/news/national/exotic-animal-owner-killed-by-650-pound-tiger/article4301532/?ref=http://www.theglobeandmail.com& Website visited 29 September 2017

Humane Society. 20 April 2016 Which cat is living next door? http://www.humanesociety.org/news/magazines/2016/05-06/exotic-pets.html?credit=web_id85539248 Website visited 29 September 2017

Epilogue

BBC News. 24 February 2011. Fox lived in the Shard skyscraper at London Bridge. https://www.bbc.co.uk/news/uk-england-london-12573364 Website visited 12 December 2018

Spiegel. 10 June 2009. Thieving fox amasses 120 shoes. http://www.spiegel.de/international/zeitgeist/imelda-strikes-again-thieving-fox-amasses-120-shoes-a-629778.html

Evening Standard. 18 April 2016. Fox boards London bus and gets off at Imperial War Museum. https://www.standard.co.uk/news/london/fox-boards-london-bus-and-gets-off-at-imperial-war-museum-a3227556.html Website visited 12 December 2018

BBC News. 24 November 2017. Fox hitches free ride on sightseeing bus. https://www.bbc.co.uk/news/uk-england-london-42112907

Herald Sun. 8 May 2018. Fox snapped riding route 57 tram at West Maribyrnong. https://www.heraldsun.com.au/news/victoria/fox-snapped-riding-route-57-tram-at-west-maribyrnong/news-story/7a6d25cdd5babeee591c59d6fd33c6ab Website visited 12 December 2018

CTV News. Undated. What does the fox dream? Critter naps on an Ottawa bus. https://www.ctvnews.ca/video?clipId=404976&playlistId=1.1928025&binId=1.810401&playlistPageNum=1&hootPostID=d636fe76c9052049098ddfd0244ba5be Website visited 12 December 2018

IMAGE CREDITS

All photographs courtesy of the author, except:

p. 20 Red fox skulls and jaw and jaws from Ightham Fissure, near Maidstone. Plate III in Sidney H. Reynolds, *A Monograph of the British Pleistocene Mammalia. Vol. II, Part III: The Canidae*, London, 1909 (Wikimedia Commons)

p. 24 *Prince Hanzoku Terrorised by a Nine-Tailed Fox*, Utagawa Kuniyoshi (1798–1861) (Wikimedia Commons)

p. 52 Black fox painted by John Audubon, from *The Viviparous Quadrupeds of North America* / by John James Audubon, F.R.S. &c. &c. and the Revd. John Bachman, D.D. &c. &c. (1847) (William L. Clements Library, University of Michigan Special Collections Research Center)

p. 64 Fox cub aged three to four weeks (25 May 2007, by Hans-Jörg Hellwig, Wikimedia Commons CC BY-SA 3.0)

p. 76 Mediterranean hackberry (from Prof. Dr. Otto Wilhelm Thomé, Flora von Deutschland, Österreich und der Schweiz 1885, Gera, Germany, Wikimedia Commons)

p. 80 Short-tailed field vole (11 December 2006 by Fer boei at Dutch Wikipedia, Wikimedia Commons CC BY-SA 3.0)

p. 131 Fox with mange (3 December 2011 by Juan Iacruz, Wikimedia Commons CC BY 3.0)